Today, every CEO strives to bridge the gap between the promise of the Generative Artificial Intelligence and the practical way to implement GenAI in their organization that will enhance productivity and create value. Ashwin Mittal and Mohanbir Sawhney address this challenge, proposing a framework that is both rigorous and realistic. A must-read for every leader thinking about AI adoption, i.e. every business leader!
Sergei Guriev – Dean of London Business School and former Chief Economist of the European Bank for Reconstruction and Development

A smart, practical roadmap for leaders who want to move from AI experimentation to real enterprise impact.
Liat Ben-Zur – Former Corporate Vice President, Microsoft; Board Director and AI Transformation Advisor, LBZ Advisory LLC

Artificial intelligence has enormous potential, but many organizations struggle to translate that potential into meaningful business outcomes. Drawing on Ashwin Mittal's decades of experience leading AI transformations at global companies and Mohanbir Sawhney's deep expertise in innovation strategy, this book provides a thoughtful and practical framework for bridging that gap. The AI Impact Model offers leaders a structured way to evaluate, prioritize, and scale AI initiatives with confidence.
Ronnie Screwvala – Co-Founder of upGrad, Swades Foundation, and UTV

Organizations across industries are investing in artificial intelligence, yet many struggle to move beyond isolated pilots. This book clearly explains why and provides a practical, outcome-oriented framework to overcome those barriers. By combining practitioner insight with academic rigor, Mittal and Sawhney offer leaders a credible roadmap for turning AI initiatives into sustainable competitive advantage.
Anupam Mittal – Founder of People Group (Shaadi.com) and Shark, Shark Tank India

I've seen firsthand that we are not short of AI ideas; we are short of AI outcomes. What stands out in this book by Ashwin Mittal, Chairman of C5i, and Prof. Mohanbir Sawhney is its focus on execution and value creation. It gives leaders a clear, structured way to move from pilots to real, measurable business impact.
Ram Iyer – General Manager, Windows Devices & Sales, Marketing, Microsoft

Many organizations invest heavily in AI but struggle to see real returns. This book provides a structured and highly practical approach for evaluating and scaling AI initiatives that truly matter.
Suraj Srinivasan – Philip J. Stomberg Professor of Business Administration, Harvard Business School

This book addresses one of the most pressing questions facing business leaders today: how to ensure AI investments deliver measurable returns. Mittal's practitioner perspective, shaped by years of guiding Fortune 500 organizations through AI adoption, combined with Sawhney's research and academic rigor, results in a highly practical guide. Leaders will find a clear framework for evaluating opportunities, avoiding common pitfalls, and building AI capabilities that generate real value.
Ajit Sivadasan – Ex VP and GM, Global eCommerce, Lenovo

Artificial intelligence is reshaping industries, yet translating AI initiatives into consistent business impact remains a challenge for many organizations. Drawing on decades of applied AI, consulting, and academic experience, the authors present a disciplined framework for evaluating, prioritizing, and scaling AI initiatives. An essential guide for leaders focused on sustainable value creation.
Deepak Jose – Global Data, Analytics and AI Executive.
Vice President, Head of Data and Decision Intelligence, Niagara Bottling; formerly served in executive roles at Coca-Cola and Mars

This book stands out for its clarity and practicality. Rather than offering abstract ideas about AI, the authors provide a structured framework for understanding where AI can create real value and how organizations can scale successful initiatives. Leaders looking for a disciplined approach to AI strategy will find this book highly valuable.
Doug Hillary – Strategic Advisor, C5i, former SVP of Analytics at Dell

What distinguishes this book is its ability to frame AI as a leadership and organizational challenge, not merely a technical one. It underscores the importance of alignment, decision-making, and accountability in driving meaningful outcomes. That perspective makes it especially relevant for senior leaders.
Alex Melen – Co-Founder, SmartSites

This book encourages leaders to ask better, more strategic questions. That alone makes it a valuable contribution.
Shawn Johal – Business Growth Coach, Elevation Leaders, Bestselling Author of *The Happy Leader*

THE RETURN ON AI

From Promise to Profit in the Age of Intelligent Business

ASHWIN MITTAL
MOHANBIR SAWHNEY

Published by:
Course5 Intelligence Inc.
3600 136th Pl SE
Bellevue, WA 98006, USA

ISBN 979-8-9956016-0-9 (paperback)
ISBN 979-8-9956016-9-2 (hardcover)
ISBN 979-8-9956016-8-5 (ebook)

Library of Congress Control Number: 2026909759

Dedication

Ashwin Mittal

To my three children, Ria, Pari, and Dev. My constant source of inspiration, perspective, and purpose.

Mohanbir Sawhney

To Shalini–my partner and my muse, who gives me the space to think, the inspiration to create, and the anchor to keep me honest.

Table of Contents

Disclaimer

All industry leaders and contributors quoted herein have shared their views in their personal capacity based on their professional experience. These views are theirs alone and do not necessarily reflect the official positions or policies of their current or past employer organizations, nor do they imply an endorsement of this book or any related commercial offerings by said organizations. Professional roles and organizational affiliations are stated as of the time of finalization of this manuscript and are included for illustrative purposes to establish the contributor's relevant expertise.

Company and organization names referenced in this work do not imply any partnership, sponsorship, or endorsement by these organizations.

Acknowledgements

The authors would like to thank our editor, Joshua Iversen, for his skilled assistance throughout the writing of this book.

We would like to thank the many C5i team members who have contributed to the building of the AI Impact Model described in this book, and also many of our Client Advisory Board members who have advised on it. In that respect, from the C5i team, we would like to specifically mention Manish Srivastava, Anil Damodran, and Paresh Banka (now ex-C5i) for the role they have played. We would also like to thank Manish Srivastava and Megha Chaudhry from the C5i team for the extensive support they have provided to the authors in completing the manuscript.

We are also deeply grateful to the industry leaders who shared their time and insights with us. Their contributions, shared in their personal capacities, have added immense value to this project. The quotations they have provided on various aspects of AI planning and implementation are found throughout the book. They include, in alphabetical order, Daisy Chittilapilly, Manik Gupta, Doug Hillary, Ram Iyer, Deepak Jose, Ujjwal Sehgal, and Ajit Sivadasan.

Foreword

Philip Kotler, S.C. Johnson & Son Distinguished Professor of International Marketing, Kellogg School of Management, Northwestern University (emeritus)

Companies today are investing huge amounts of capital into artificial intelligence initiatives. They can't take the chance of being weak in AI.

Yet they have many questions to answer:

- Which AI projects should they invest in and which should they pass on?
- Is there a way to estimate a project's potential success prior to expending resources on it?
- What factors should be analyzed when evaluating the potential return on AI that a particular project is likely to generate?
- How can successful AI projects be effectively scaled?

I am most impressed with this new book by Ashwin Mittal and Professor Mohanbir Sawhney. Their book offers incisive answers to these questions without mitigating the complexities that companies face in productively implementing AI.

A most helpful contribution is their innovative AI Impact Model. This model is designed to predict and optimize the success of AI initiatives in order to enable organizations to improve their ability to generate satisfactory return on AI. The model uses five Value Levers–Problem, Data, Technology, Talent, and Execution. The company must define its desired problem outcomes, gather quality data, use the right technology, recruit the right talent, and execute smartly.

The book identifies the Reality Gap that is often seen between project objectives and results as a major source of frustration among organizations implementing AI. To provide a nuanced depiction of why AI projects succeed or fail, it examines the intricacies involved in the interactions between the five Value Levers. Additionally, it contains a detailed description of the use of simulation to pretest different scenarios pertinent to particular AI projects. In the final chapters, the authors detail plans for further improvements to the AI Impact Model and insightfully delve into potential paths for the future development of AI and how this may affect the model.

The Return on AI: From Promise to Profit in the Age of Intelligent Business is highly readable and practical. It includes many examples of AI use cases, including successful implementations as well as projects that illustrate some of the problems that can occur when applying AI.

Today's business leaders face immense pressure to invest in AI as a competitive necessity. All too frequently, they find themselves navigating without a chart. They need a reliable framework to evaluate and prioritize their AI initiatives. The AI Impact Model detailed in this book constitutes a powerful tool that business leaders, decision-makers, and AI practitioners striving to maximize their returns on this transformative technology can employ to select and optimize AI projects.

Preface

Ashwin Mittal

The genesis of the model designed to predict the success of AI projects and prescribe action to improve their impact presented in this book stemmed from an exercise aimed at redrafting our company's value proposition.

At C5i, we empower our clients by transforming their decision-making. Utilizing cutting-edge AI, data science, and analytical methodologies, we deliver bespoke solutions for global corporations. We work across various business functions in our client organizations, including sales and marketing, customer experience, omnichannel commerce, supply chain, human resources, and enterprise technology, to scale AI across the organization. We utilize our enabling expertise around data and AI science, along with a wide library of our own AI products and accelerators, to provide consulting and solutions to many leading companies around the world.

When we went about defining our value proposition, we realized that we were all very clear that we did not want it to be about being the best technology or AI provider for the sake of technology or AI. Our focus is to be a true partner to our clients, and our value proposition is all about driving meaningful business impact for the enterprise. That is what we are passionate about and stand for.

However, we do realize that often there is a gap in making business impact a reality in practice, which is not in our control. At the core, we are a science company. Our focus is to understand the business problem, build AI algorithms from the data, and then build a solution that can be deployed by the enterprise. But there are a lot of things that happen within these large complex organizations of the clients we work with when they implement AI; these could be centered on data or other factors such as people, change management readiness, culture, etc.

We may build the AI algorithms perfectly, but if the initial problem itself was not defined appropriately, or good data was not available, or the organizational ecosystem is not amenable to operationalization of that algorithm, the desired results will not arrive. Companies often invest heavily and end up with "graveyards" of projects that have not been implemented at all or have been discarded after initial implementation. Thus, we find that while some clients and some AI programs achieve and even exceed business goals, in quite a few cases, the outcomes fall short of expectations.

The specific ideation of the model designed to help enterprises address this issue and thereby increase their return on AI investments occurred in New Jersey in 2022. I was there for our upcoming customer advisory board meeting being held in Cape Cod, Massachusetts. I happened to be in New Jersey beforehand to meet some of our key senior customers and stakeholders. It was late at night and, suffering from jet lag, I was unable to get to sleep. Instead of reading a book or working through my email, I started thinking about our promise to help our customers achieve positive business impact. With nothing to distract me, I had the luxury of plenty of time to consider the topic. Specifically, I asked myself the following questions: "Are we doing enough to make that real— to deliver the necessary business impact? And what else can we do to enhance the impact of our work for clients?"

This train of thought led me to the question which, eventually, resulted in the development of the AI Impact Model (AIIM) covered in this book. After pondering how to maximize the delivery of business impact for a time, I further asked myself: "If we are a science company that is helping companies predict outcomes in their business, such as which customers are likely to buy what, which marketing is likely to be effective, etc., shouldn't we use scientific techniques to deliver value from AI?" After all, I thought, "We are building science to help companies predict outcomes. But then, when they employ our science in their business, they don't always get the predicted business impact they expected from that science. To improve results, can we build science on our own science to help them predict and optimize the business impact of AI before they use it? Can we build AI for AI?" That was the question that came to my mind, and that drove me to sketch out the broad model of the AIIM.

After putting an outline of the model together, I showed it to my colleagues. They were immediately intrigued by the idea. Then, we said, "Let's present this at our customer advisory board meeting and elicit some feedback from customers." In doing so, we received a wealth of valuable feedback and many "ahas," demonstrating significant interest in the concept. After that, the idea kept germinating, although we didn't end up implementing it right away. At some point during that process, we met one of our customers from a large global enterprise who was visiting us in India. I showed him the idea, and he said, "This is phenomenal; we must build it together."

Then I showed it to our advisor, Professor Sawhney. He said, "This is very interesting, but maybe you can think about this a little differently." His feedback helped us refine the thought process. Then, we partnered with various of our customers–leading Fortune 500 and Forbes 2000 global enterprises–to further develop the AIIM. That's the story of the inception of the model which forms the center of this book, which is all about how AI can be utilized to drive meaningful business impact and transform organizations.

Prof. Mohanbir Sawhney

About twenty years ago, I built a framework called the Innovation Radar and wrote an article explaining it. In devel-oping the framework, we looked at the different types of inno-vation that companies typically use and how to measure the impact of innovation. We came up with twelve dimensions in a type of circular radar framework. Innovation could be plotted as stemming from any combination of these factors.

When Ashwin called me and told me he was working on this AI Impact Model, I said, "Instead of just having a laundry list of variables, we should think about some sort of a visual construct of this impact model." That led to the spider chart demonstrating how the model we'll cover in this book works. After that, we worked together on this notion of "Why not use AI to predict AI?" This was the meta concept of "Why not build an AI model to predict the success of AI models?"

The concept Ashwin brought to me was: how can we best understand this idea of impact, a problem many managers face? He had this notion that we could use data science to help with this, and I helped him structure it into the five main factors and eighteen subdimensions which make up the model. It became a melding of the minds where what I brought to the table were frameworks and structure to use with the idea he presented. Our collaboration was assisted by my work as an advisor to Ashwin's company for more than ten years, which gave me a solid understanding of what would work best given the company's capabilities.

Our complementary skill sets and perspectives included Ashwin's domain expertise and client expertise, combined with my contribution of academic frameworks and rigor. This book takes those starting points and demonstrates the value proposition of the AIIM as rooted in practice, with real clients and real data, but at the same time informed by a rigorous academic framework.

To accomplish this, I took some of the ideas from my paper on innovation and brought them over for use in developing this framework. In the following chapters, we'll provide a detailed account of the model that resulted from our collaboration and how it can be utilized to enable enterprises to more effectively deploy AI.

Introduction

Why This Book?

Artificial intelligence is a transformative technology that promises to create enormous business value by automating routine tasks, delivering personalized customer experiences, and unlocking new revenue streams. The potential business impact of AI is massive, but this potential is not being realized in practice. Most AI initiatives fail to deliver significant business value. According to a recent MIT study, ninety-five percent of generative AI projects failed to deliver measurable returns on investment.[1]

In today's information-saturated world, the age of intelligent business, this disconnect between promise and profit is a significant problem for organizations. Business leaders face immense pressure to invest in AI as a competitive necessity, but they are often flying blind, with no clear framework to evaluate, prioritize, or optimize their AI initiatives. They need answers to questions like:

- Which AI projects are worth investing in?
- How can we predict the likelihood of success before committing resources?
- What factors determine the outcomes of our AI initiatives?
- How do we scale successful AI projects across the organization?

Without clear answers to these questions, organizations risk wasting resources on poorly defined projects, getting stuck in endless pilot projects, and failing to scale their efforts. Several hurdles stand in the way of realizing business value from

AI–unclear problem definitions, poor data quality, inadequate technology, and a lack of skilled talent. Yet organizations lack a structured approach to harnessing the potential of this transformative technology to deliver the return on AI they expect.

The Return on AI: From Promise to Profit in the Age of Intelligent Business addresses these pain points by introducing the AI Impact Model (AIIM)–a practical, evidence-based framework that helps organizations measure, predict, and optimize the outcomes of their AI initiatives. The AIIM framework is designed to:

- **Predict AI Success**: The AIIM provides a systematic approach to assess the likelihood of success for any given AI project, based on five factors called "Value Levers": Problem, Data, Technology, Talent, and Execution. By evaluating these factors, business leaders can make data-driven decisions about which projects to prioritize and where to invest resources.

- **Optimize Outcomes**: Beyond prediction, the AIIM framework enables organizations to improve the success rates of their AI initiatives by identifying specific areas for intervention, such as improving data quality, upskilling talent, or refining execution strategies.

- **Scale Success**: A challenge with AI adoption is to move from isolated pilot projects to enterprise-wide implementation. Scaling challenges are most often related to organizational readiness. The AIIM framework provides a roadmap for scaling AI initiatives and planning related investments, ensuring they deliver consistent and sustainable value across the organization.

The problem we address is pervasive and profound: ***how can organizations quantify and improve their return on AI***

investments, while avoiding the common pitfalls that lead to failure? Unlike other AI books that focus on technology or algorithms, we emphasize the business impact of AI. This book is not a technical manual but a practical guide for business leaders, decision-makers, and AI practitioners who need a structured way to evaluate and optimize their AI initiatives.

By combining academic rigor with real-world insights, the authors equip readers with the tools and knowledge they need to navigate the complexities of AI implementation. Through case studies, data-driven insights, and practical recommendations, the book provides a roadmap for achieving meaningful, measurable, and scalable results from AI. We have also interviewed several AI thought leaders from leading global organizations who have provided quotes on the subject, which are featured throughout the book.

This book is for every organization that has struggled to move beyond AI pilots, for every business leader frustrated by the lack of tangible results from AI investments, and for every practitioner seeking a clear framework to guide their work. *The Return on AI: From Promise to Profit in the Age of Intelligent Business* is the solution to the challenges organizations face in their AI journeys. It is a guide for those who believe in AI's potential and are ready to unlock it.

How This Book Is Different

This book distinguishes itself through several key differentiators:

Empirically Validated, Data-Driven Framework: Unlike many works that offer conceptual discussions, this book introduces the AI Impact Model, an empirically validated framework derived from an in-depth analysis of sixty-two AI projects across various industries. This data-driven approach ensures that the insights and recommendations are grounded in real-world evidence, providing readers with actionable strategies to predict, measure, and optimize the outcomes of their AI initiatives.

Complementary Expertise of the Authors: The collaboration between Ashwin Mittal, Chairman and Co-Founder of C5i, and Prof. Mohanbir Sawhney of Northwestern University's Kellogg School of Management, brings together a unique blend of practitioner experience and academic rigor. Mittal's extensive background in guiding Fortune 500 companies through AI-led transformation complements Sawhney's scholarly research and advisory roles in technology and innovation. This partnership ensures that the book balances theoretical foundations with practical applications, making it both insightful and immediately applicable.

Comprehensive Coverage of Success Factors: The AIIM framework is holistic in its approach, encompassing five key Value Levers: Problem, Data, Technology, Talent, and Execution. By addressing both technical and organizational dimensions, the model provides a thorough roadmap for AI project success, ensuring that no essential aspect is overlooked.

Predictive and Prescriptive Simulation Tool: A standout feature of this book is the inclusion of a simulation tool based on the AIIM framework. This tool enables organizations to forecast the potential success of AI projects before significant resource allocation and to optimize outcomes by modeling various scenarios. By adjusting factors such as data quality or execution strategies, leaders can identify the most impactful interventions, facilitating informed, data-driven decision-making.

Proven Real-World Impact: The AIIM framework has been successfully applied in real-world settings, delivering tangible benefits to various organizations. For instance, one of our clients in the technology industry utilized the model to optimize its AI project portfolio, achieving enhanced efficiency and business value. Similarly, the same company leveraged the framework to align its AI strategy with organizational objectives, resulting in measurable success across multiple initiatives. Another client in the consumer goods industry uses the AI Impact Model to evaluate and assess the adoption of its AI initiatives.

These case studies underscore the practical effectiveness of the AIIM framework in driving meaningful business outcomes.

In addition to case studies of actual AI implementations, the book includes "scenarios" that draw on the authors' extensive experience in the sector to describe hypothetical accounts of AI use cases, both successful and unsuccessful.

While several notable books have explored the intersection of AI and business, *The Return on AI: From Promise to Profit in the Age of Intelligent Business* sets itself apart through its empirical foundation, the complementary expertise of its authors, a comprehensive and actionable framework, innovative predictive tools, and demonstrated success in real-world applications.

Who This Book Is For

This book is designed for a diverse audience of leaders, practitioners, and innovators across industries. It speaks directly to those tasked with driving artificial intelligence initiatives and offers a clear roadmap to ensure success. Whether you are an executive, an entrepreneur, a consultant, an AI practitioner, or a curious academic, this book provides actionable insights that will resonate with your unique challenges and goals.

Business Executives and Decision-Makers: For C-suite leaders, general managers, and functional heads, this book offers a framework for navigating one of the most pressing questions in today's business environment: how can we unlock measurable value from our AI investments? These leaders are tasked with allocating resources, making high-stakes decisions, and ensuring that AI initiatives deliver real returns. The AI Impact Model equips them with a tool to assess the likelihood of project success, prioritize investments, and avoid costly failures.

AI Leaders and Practitioners: For Chief Data Officers, AI project managers, data scientists, and machine learning engineers, this book provides a structured, systematic approach to evaluating and optimizing AI initiatives. While the technical community is often proficient in the mechanics of AI, the challenge lies in connecting their work to business outcomes. The AIIM framework bridges this gap by offering a holistic view of success factors–from problem definition to execution excellence. Practitioners will gain practical tools to communicate the value of their projects to non-technical stakeholders and to identify critical areas for improvement in their initiatives.

Consultants and Advisors: Consultants, strategy advisors, and business analysts play a critical role in guiding organizations through digital transformation. This book is an essential resource for those advising clients on how to implement AI effectively. The AIIM framework offers a repeatable, evidence-based methodology that consultants can apply across industries and organizational contexts. By using this framework, advisors can help their clients make informed decisions, navigate trade-offs, and optimize their AI investments for maximum impact. Additionally, the included simulation tool provides consultants with a powerful resource to model outcomes and recommend targeted interventions.

Academics and Students: For academics researching AI or students aspiring to lead in AI-driven fields, this book serves as a practical complement to theoretical knowledge. While much of academia focuses on algorithmic development and technical advancements, *The Return on AI: From Promise to Profit in the Age of Intelligent Business* focuses on application and impact. By exploring the organizational, cultural, and strategic dimensions of AI, the book fills a critical gap, offering insights that are often missing in purely technical discussions. Students and researchers will gain a deeper understanding of the business context for AI and learn how to bridge the gap between technology and value creation.

This book is a guide for every professional who wants to make AI a driver of measurable and meaningful impact. By addressing the diverse needs of its audience, the book ensures that readers at all levels and from all disciplines walk away with actionable insights, practical tools, and the confidence to lead in an AI-powered world.

How the Book Is Organized

This book is structured as a journey that takes readers from understanding the challenges of AI adoption to mastering a comprehensive framework for delivering measurable business value. After analyzing the difficulty organizations often experience in generating return on AI in Chapter One, the book introduces the AI Impact Model *(patent application pending at the time of writing/publication)* in Chapter Two. Chapters Three through Seven gradually introduce the Value Levers which, along with their sub-levers, are responsible for AI project success or failure. Analysis of these factors is enhanced by illustration of their practical application through real-world examples and hypothetical scenarios based on the authors' extensive experience in the sector.

The narrative blends data-driven insights with actionable recommendations, taking an approach that is aimed at being both engaging and deeply practical. Chapter Eight examines the impact of the interaction of the factors involved in AI project success. The final three chapters of the book cover the model's simulation process for gauging and enhancing the potential success of AI projects as well as offering a look into the future of the model and AI in general with reference to the latest developments in the sector, including generative and agentic AI.

CHAPTER ONE

The AI Promise and the Reality Gap

Imagine that a corporate CIO is charged by the company's CEO with pursuing an aggressive AI implementation program. He consults with experts inside and outside the organization and draws up a list of fifty AI projects. Twenty of them are chosen for implementation, and of those, roughly twenty-five percent prove to be effective, given the time and resources invested in them. The resulting return on investment (ROI) of this AI implementation program, after all the expenses and the improved efficiency of the successful projects have been accounted for, is net negative. To the dismay of the CIO, despite some successful projects, the company has spent more money and management attention on implementing these projects than it expects to realize in improved operational efficiency from them over a reasonable time frame.

Unsurprisingly, the CEO is not pleased by these results. She tells the CIO to implement another round of AI projects, but this time, he is instructed to be more selective and only choose the most promising projects for implementation. Following her instructions, the CIO once again compiles a list of fifty potential AI projects from a variety of sources. This time, instead of implementing twenty of them, he selects the ten he feels are most promising. To his dismay, in this round, only two, or twenty percent, of the projects achieve success. The results for this group of projects are even worse from an ROI standpoint!

Fearing that the CEO will abandon AI adoption and the company will see its competitiveness erode as a result, the

CIO requests permission to launch one more attempt to achieve a positive business impact from AI. Reluctantly, the CEO agrees, but only after making sure the CIO understands how high the stakes are. The company's Board of Directors is watching closely, and another failure will place the company at a disadvantage vis-à-vis other industry participants that have successfully integrated AI into their operations. The CIO gets the message. He knows he needs to get it right this time, or the company's status as an industry leader, and perhaps even his own position, could be at risk.

What steps could our hypothetical CIO take to get the desired business impact and a favorable return on AI on his third attempt? The factors that he should take into account, and an approach he can use to simulate how to most effectively employ AI at his enterprise, can be found in the contents of this book. Adopting such a strategy centers on taking a scientific approach by modeling projected AI project performance based on factors instrumental to AI success, such as problem definition, data, technology, talent, and execution.

The use of these AI success factors, or Value Levers, lies at the heart of the AI Impact Model (AIIM) featured in this book. As covered in the Preface, what motivated us to develop this model in the first place was the goal of using it to help enterprises and individuals effectively generate business impact from AI by solving issues such as those facing the stressed-out CIO in the scenario above.

In the following chapters, we'll explore exactly how the model works and provide more details of how the factors mentioned above–and their sub-factors–can be used to evaluate and enhance a project's potential. For the remainder of this chapter, we'll explore the various challenges and issues organizations face when attempting to implement AI projects. Failing to overcome these challenges explains why, as in the scenario above, many enterprises experience difficulty when attempting to realize the ROI they expect from AI projects.

AI Project Deployment: Expectations versus Reality

The tremendous increase in computer processing power in recent years has seen artificial intelligence move from a largely theoretical application to one capable of transforming organizations throughout the business world. Yet, despite billions of dollars in investments, most AI projects either end up mired in the pilot process or fail to scale. This underperformance, this failure to move from promise to profit, means that the transformative potential of AI often falls short in reality, making it difficult to predict the business value it is likely to deliver.

As enterprises strive to adopt AI across their ecosystems, business leaders are focused on trying to measure, predict, and optimize the return on AI projects. The challenges enterprises face in determining the business impact of AI can make it difficult for them to realize value at scale when implementing these projects. In addition to the study by MIT highlighted in the Introduction, which found that ninety-five percent of GenAI initiatives fail to deliver returns, this is also supported by a KPMG survey of US business leaders, which discovered that only fifteen percent of those responding to the poll reported having established metrics they could use to measure returns on their generative AI investments.[2]

Some of the most significant challenges to delivering business impact from AI include data quality issues, resistance to change, lack of clear objectives, lack of skilled AI professionals, inability to integrate AI into workflows, and inadequate data governance. Overcoming these hurdles requires an understanding of the key factors that drive AI success. Evaluating these factors prior to embarking on an AI initiative enables business leaders to improve their return on AI investments by focusing on those projects most likely to be successful. Adopting this approach enables organizations to optimize their AI strategy, helping them maximize returns on the deployment of AI across the organization.

In our experience, a major issue plaguing AI implementation is the lack of a reliable approach to determining which projects are likely to be successful. It's easy enough to write a theoretical project map positing hypothetical impacts; however, without real, structured analysis backing up these hypotheses, it's hard to know how likely it is that they will deliver the desired results when implemented. Carrying out such an analysis prior to launching an AI project, in our view, is key to boosting the likelihood of success. This is where the rubber meets the road.

Failing to Take a Holistic Perspective

A major reason companies fail to prioritize the most promising AI projects revolves around the failure to take a truly holistic perspective. In this regard, we could cite the parable of the six blind men and the elephant, where each individual focuses on only a part of the elephant (tusks, legs, tail, trunk, etc.) and thus the group doesn't realize what they are really dealing with. In the parable, one of the men grabs the trunk and says, "This is a hose; therefore, this must be a fire truck." Another, feeling the side of the elephant's body, says, "This is a wall, so it must be a building," while another, feeling a leg, says, "This is a trunk, so it must be a tree."

Organizations can make the same type of mistake when preparing AI projects. The problems they often encounter are an artifact of the part of the organization or the point of view that's taking the lead. For instance, if an enterprise's IT department leads the project and serves as its primary driver, it can risk losing sight of the use case–the problem definition. The reason is that they're not getting enough engagement with business stakeholders from other departments, such as sales and marketing, supply chain, HR, etc. On the other hand, if business stakeholders are driving AI adoption without much support from the IT team, maybe they didn't use the right tools and don't think enough about the data in deploying the project.

The takeaway is that to do a true AI-driven transformation, it's important to think systemically and holistically across the entire enterprise. That type of thinking can be rare because we are all biased by our own silo that we sit in. This is often true even at the C-suite level. If you're a CIO, for example, you look at things differently, through a tech lens. If you're the CFO, you're looking at the world through a financial lens. As a result, it becomes challenging to put all these pieces together. You can think of this as a three-legged stool. If one of the legs becomes weak, the stool wobbles–it doesn't work properly.

AI projects can be like this three-legged stool, where each leg is set to be installed by a different person. But then the project must sit on that stool; if one leg is shorter or one is weaker, it makes things unstable. The systemic point to draw from this is that unless all your legs–all aspects of your AI project–are strong and level or equal, you're not going to have a strong stool; certainly not one that you can count on to support weight or–for our purposes–a successful AI implementation.

This is the silo problem or the functional bias problem–essentially, a point of view that is not looking at the whole picture. For instance, as with the prior example of a company CIO, or any tech leader, if a person's expertise is in technology, it can be natural to have the tendency to focus on technology and tools. If you're in HR, the tendency is to focus on people and skills. If you're a business leader, you'll typically focus on the problem and the use case. But you need to take all of these aspects of the project into account to achieve a successful implementation.

Talent and Technology

Another challenge of successfully implementing AI is that companies may lack the expertise or experience to optimize their use of the technology. This generally happens with any new technology–it takes a while to diffuse, evolve, and get adopted. As a result, getting the most out of AI takes a level of maturity and understanding amongst people using these tools themselves. This gives an organization the ability to absorb, adapt, and put their AI tools to use. Typically, technology moves faster than our ability to effectively implement and utilize it. Given its relative novelty, most technology experts didn't grow up using AI. Thus, it can take some time for them to master these tools, particularly in the context of unlocking business value or scaling an implementation as opposed to running pilots or small projects, which serve as proof of concept.

Technology often jumps ahead in exciting ways, and then organizations and experts scramble to catch up. We're clearly seeing that with AI, as demonstrated by the statistic cited earlier regarding ninety-five percent of generative AI projects failing to deliver the expected business impact. While AI has seen increasing use in the corporate world over the past ten years, and in a few cases, even before that, the pace of the evolution of the technology over that period has been rapid. We've seen innovations and updates in AI functionality on a regular basis–every year, every six months, every three months, and in recent times, even every week. This rapid change challenges organizations, users, and IT experts to keep up with the latest iterations. Above and beyond this challenge is the fact that when AI projects are built, the AI algorithms or models underlying them ultimately have to be deployed in an organization's environment–in their IT ecosystem.

This can be challenging because these models are typically built in a simulated environment. However, when projects are implemented in the real world, a number of other factors apply. These factors can be hard to plan for and thus can produce results that fail to meet expectations. There are various dynamics at play within the organization itself. These all relate to the five Value Levers for deploying AI successfully, which we will discuss in more detail in the next chapter. They include identifying the problem and structuring the solution, various attributes of the data, the people skills necessary for implementing and using the AI, the technology available to deploy the AI, and project execution. When the presence of these factors is not properly optimized in an AI project, it can impede effective implementation, leading to failure to drive the anticipated return on investment.

The takeaway here is that to tap into the transformative power of AI, organizations need to take into account a variety of factors that, if not addressed, can prevent a project from being successful. While the excitement over AI's potential impact is not unwarranted, failure to adequately incorporate these factors in the planning process, in our analysis, plays a major role in unsatisfactory AI implementations.

Without the right talent to successfully deploy and utilize AI, a project's outcomes are likely to fall short. Similarly, if an organization's culture is not ready to embrace the use of AI, perhaps because team members are not accustomed to using it, the project is also likely to fail to meet its goals. Failure can also stem from a lack of quality data or from technology systems that aren't designed to adequately orchestrate an AI implementation.

Even when an organization has the proper technology, its staff may lack the training to properly use it. For all these reasons, executing an AI project can be a complicated process, and we believe this plays a large role in the gap between what organizations expect AI can achieve and the actual results they are experiencing.

Defining Expected Value

A key error we have seen organizations make is that they sometimes fail to define the expected outcome or value of a specific project. This can make it difficult to evaluate how much actual value is being captured via the implementation of an AI project. For instance, a project's parameters may fail to identify how much revenue or cost savings it is expected to generate within a certain timeline. Another measure of value added might be improvements to the customer experience, leading to increased customer loyalty. Yet another could be a reduction in employee attrition. Whatever the metric is, unless you define it and track it, determining a project's business impact will be difficult, if not impossible.

"I have seen that clear problem definition and good change management planning before starting an initiative are critical to achieving desired results. If the organizational process changes post-implementation of an AI program are not envisaged and planned for, then AI programs built with even a strong foundation of data and science can fail to achieve desired results."

Ajit Sivadasan — Ex VP and GM, Global eCommerce, Lenovo

Failure to define a value objective for a project can lead to an inability to achieve measurable success, which makes it hard to gain the momentum needed to attract investment and management focus for AI initiatives. This is why we stress to companies that there should be a value metric associated with each AI project. The simulator associated with the model for evaluating and optimizing the business impact of AI, which we will present in Chapter Nine, requires that users record a business value for each AI project they are considering. This feature stems from what experience has taught us is the importance of identifying an expected value to be realized before launching any AI project.

Publicity and AI

Another issue that can lead to AI projects failing to generate the expected business impact is the amount of excitement, and even hype, surrounding AI these days. We are in a situation where there is early support and enthusiasm for AI that neither author has seen in our combined multiple decades of work in academia and consulting. We believe that the transformative potential of

AI is real—that it can and will bring substantial change across the world of business. That said, with such transformative technologies, there is often the tendency to overestimate what can be achieved in the short term and underestimate what can be achieved in the long term.

To our knowledge, there has not been a prior technology that has been so enthusiastically pushed by the corner office. Normally, it's the IT folks trying to sell a technology to the CEO, but with AI, in many cases, it's often the board and the CEO saying, "Let's put this to use." So, it's like a shiny object that attracts attention because so many people have played with ChatGPT or a similar app by now, and almost everybody has, to some degree, experienced firsthand the power of AI.

For all these reasons, the push to launch AI programs has gained significant momentum at many enterprises. The motivation is the return in the form of business impact that AI can generate. However, the fact of the matter is that, often, the enterprises deploying these projects are not ready to derive the expected value from implementing AI. They typically fall short in a number of the five Value Levers and eighteen total sub-levers necessary to extract value from an AI deployment that will be covered in Chapter Two.

The next step is to bring AI into the enterprise. But doing so takes planning and resources—without them, the promise of AI is unlikely to be fulfilled. The good news is that AI has been democratized, so everybody can actually use these tools. The bad news might also be that everybody can use these tools. This can cause people to think that just because they can ask ChatGPT and get an answer, we should be able to put this into the enterprise. However, it's not that simple. This is where people can get misled by the high expectations stoked by all the publicity around AI. What we have found is that there is a big gulf between promise and profit. Our job in this book is to bridge that gap and show how organizations can successfully go from promise to profit, from hope to reality, from potential energy to kinetic energy, from strategy to action, outcomes, and return.

The extensive publicity around AI can also, instead of motivating enterprises to try the technology, have the opposite effect in some cases. Negative publicity around the risk of AI may dampen the enthusiasm among some organizations to invest in such projects. In this case, they may miss the opportunities for realizing business impact from AI available to enterprises that properly analyze its potential and prepare the way for its deployment.

Additionally, talking about speculative futuristic developments in AI, such as artificial general intelligence that matches or exceeds that of humans, can lead to unrealistic expectations of what AI can do currently. Then, when a project doesn't work, it can lead to an overreaction against AI in general. Even in cases where AI does have a positive return, if expectations are excessively lofty due to the publicity relating to speculative futuristic applications of the technology, it can lead to disappointment. In these cases, the deluge of publicity associated with AI can be counterproductive and lead to disinvestment in the technology. Leaders may be tempted to say, "Let's wait and see how things shake out," thereby missing the opportunity to benefit from the productive AI applications currently available.

While the publicity around AI can tend toward the sensational, such as AI creating great art or writing poems, etc., this can obscure its current ability to drive significant value creation at more practical tasks. Excessive focus on AGI (artificial general intelligence) and the so-called Singularity (the hypothetical moment when computer intelligence exceeds human intelligence) obscures AI's current ability to drive hard business value and positive ROI. The fact of the matter is that organizations must do the hard work necessary to derive significant benefits from AI projects. The potential to create significant value is there, but there are a number of factors, which this book will cover in detail, that planners of an AI project must address first.

Sports and AI–an Analogy

A sports analogy may be helpful in illustrating the importance of modeling a variety of factors related to AI project success in selecting which projects to undertake. Professional sports is a multi-billion-dollar industry globally. In the sports world, of course, player selection is crucial to the success of any sports franchise. The problem a team's ownership usually faces is that simply trading for or hiring the best players in the league is likely to be inordinately expensive, making such a strategy out of the reach of all but the wealthiest franchises. As a result, in many professional sports, substantial effort and expense are devoted to the process of analyzing and drafting or recruiting promising young talent.

While at one time this process might have been a fairly simple affair, driven by "hunches" and monitoring the performance of young players, that is no longer the case. Given the tremendous amounts of money involved, there is often a comprehensive, data-driven process involved in deciding which players to draft. In the US, the National Football League's Draft Combine serves as an example. While a player's performance in college football may help earn him an invite to the Combine, which features the most promising young players, that alone won't necessarily get him drafted. The reason is that a star college football player's performance may be driven by being on a good team as much as by individual prowess. The Combine, by testing factors such as speed, strength and agility, allows NFL teams to gather additional personalized data to take into consideration along with factors such as the player's performance in college, their character, etc.

A player's character is important because professional sports teams understand that even if they do manage to trade for, acquire, or draft the best talent in their league, raw talent alone is not sufficient to win in most team sports. Thus, even if a team could afford to hire more of the best talent than their competitors, simply filling the roster with stars does not necessarily drive success. Teams with less talent but better teamwork will often defeat more talented teams on the basis of their ability to work together as a team. Additionally, a player's ability to complement the other players on the team in terms of what strengths are already there in the other team members and what the additional player brings to complement those existing strengths and skills is an important factor in determining whether that player will help or hinder a team's prospects of winning.

While the factors driving AI project success are certainly different than those which might predict athletic success, the principle of selecting one project over another or one team combination over another on the basis of data is the same; if you have information related to a prospective project based on the past performance of similar projects, just like having a variety of relevant data for a particular team combination, you can significantly increase your chances of selecting a winner.

In Chapter Two, we'll discuss the model we have developed to perform the type of analysis needed to position an organization to unlock the value artificial intelligence can deliver: the AI Impact Model.

CHAPTER TWO

Introducing the AI Impact Model

As a Data and AI professional, it's disheartening to see several digital transformation initiatives fall short of their potential due to challenges like organizational adoption, culture, program complexities, data issues, and cross-functional partnerships. Despite immense effort, such setbacks can be frustrating. Personally, I enjoyed working with C5i, under Ashwin Mittal's visionary leadership and in collaboration with Dr. Mohanbir Sawhney as part of the development and brainstorming of this groundbreaking AI Impact Model. This innovative framework enables organizations to assess the potential impact of AI initiatives and proactively recommend actions to enhance success across critical dimensions. I have leveraged this model during my professional journey and am proud to have contributed to its development while being an early adopter. I truly believe this initiative empowers organizations to make more meaningful, strategic AI investments and accelerate their journey toward impactful innovation.

Deepak Jose – Global Data, Analytics and AI Executive. Vice President, Head of Data and Decision Intelligence, Niagara Bottling; formerly served in executive roles at Coca-Cola and Mars

When analyzing an AI project, it's important to consider both the characteristics of the plan and the business problem it is trying to solve. To successfully execute an AI project, skilled AI scientists must parse data and generate AI models. This much is clear. But what if we were able to utilize the features which characterize an AI project to predict that project's likely business impact?

What if we could find a way to employ an AI model to evaluate the potential of AI projects?

As covered in the Preface, seeing the need to create a method of determining which AI projects are most likely to succeed, the authors collaborated on building a Model designed to use AI to predict AI. This ties in with the mission of C5i, the company Ashwin leads, which is to drive business impact through leveraging data and AI. The company focuses on using data to predict business outcomes for its clients and prescribe steps they can take to enhance those outcomes. This is done through a combination of consulting, cross-functional teams of AI practitioners, and a suite of products and accelerators.

After deciding to take on the challenge of using AI to help predict the impact of an AI initiative for a client before it is undertaken, the authors collaborated with the goal of using this process to enhance business impact. This led to the following query: "What are the questions that we would like to be able to solve for companies that will really add value?"

One question for a client looking to undertake an AI program involves asking, even before they launch the project, "Can we predict the likely outcome–what the business impact will be? Will it be in line with what we estimated? Will it be more? Will it be less?"

We've seen that our clients often have a large list of different programs they're considering, in some cases hundreds; that list needs to be prioritized because they have limited resources, limited time, limited attention, and they really want to pick those projects that will deliver the best outcome. So, can we help them prioritize? And then, if a certain program is not predicted to have the right impact, can we prescribe changes to fix that?

The creation of the model presented in this chapter, the AI Impact Model (AIIM), was inspired by the need to help C5i's clients who are deploying AI projects answer the above questions and more, including: how do we go about predicting likely business impact from AI programs before they are undertaken? How can we measure projected ROI? How do we prioritize different AI programs? Which types of AI deployments are most likely to be effective? What related organizational initiatives can the company invest in to make the AI program more successful?

The model we developed was designed to answer these questions in order to help organizations decide where to put their money when launching AI-related projects. For the model to accomplish its objective, we realized that it would be crucial to understand why AI projects succeed or fail–to identify the underlying levers that are responsible for driving their results. And that's exactly what the AIIM is trying to solve for. The goal is to drive greater impact and ROI from AI initiatives.

How the Model Works

The AI Impact Model ingests project feature information which is then used to predict the level of success a project is likely to achieve. The factors utilized by the model are called "AI Value Levers." Each of these five Value Levers is further divided into a total of eighteen sub-levers that provide additional granularity when evaluating an AI project. This is a causal model designed to predict the business impact of an AI project. The AI Value Levers serve as independent variables in the calculation process, while business impact is the outcome variable.

The model is calibrated on training data taken from sixty-two AI projects from several enterprise partners. Throughout the book, we will discuss insights taken from the model's predictions. Among these are identifying the most influential causal factors and the interactions that occur among and between the causal factors that play a part in determining business impact. The AIIM model can also be utilized for scenario analysis and simulation. This enables enterprises to project the likely improvement in business impact an AI project will deliver if investments are made in improving one or more of the AI Value Levers. In a later chapter, we'll present our simulation tool, which is designed to predict, diagnose, and optimize the business impact of an AI project.

When we analyze the eighteen sub-levers that influence AI project success, it's not just that each of them has an impact. They also interact with each other. If you have a problem that has risks centering on, for instance, data with regard to privacy or regulation, perhaps you need more technical expertise. There are various other linkages between each one of these factors. The projects selected to provide data for the model represent a mix of industries across various companies. They cover the entire spectrum of business functions including customer experience, marketing, supply chain, HR, etc.

With regard to business impact, this outcome is the variable that represents the goal of the project. The five factors, or Value Levers, involved at a high level include the problem we're looking to solve, the data, the technology, the talent, and the execution. The interplay between these five factors determines whether a project is successful. The assumption is that the functional work on executing the project has been done correctly. As long as that work has been done correctly and well, realizing the outcome is a function of these five factors and their sub-levers. We'll cover both the Value Levers and their sub-levers, or subdimensions, in the next section.

Value Levers and Sub-levers

The five Value Levers proposed by the AIIM model that influence an AI project's business impact (BI) can be summarized as follows:

The ***Problem*** (P). This lever focuses on various attributes of the business challenge and the strategic alignment of the leadership team around the business problem. The business problem has been identified as a critical success factor in studies of digital transformation initiatives.[3]

Data (D). This includes data quality, availability, and governance. Data quality has been consistently identified as a fundamental prerequisite for AI success.[4]

Technology (T). This includes the technical infrastructure, tools, and architecture required to support AI solutions effectively.

Talent (Ta). This focuses on the technical expertise and domain knowledge of the people involved with the project.

Execution (E). This measures the organization's ability to implement and manage AI initiatives effectively, including project management capabilities and operational excellence.

The AIIM model includes the most vital elements of successful AI implementation. These elements, when used properly, facilitate value creation. Each of these factors is cited consistently in literature pertaining to AI project success as vital aspects of a project's outcome.[5] Considered as a whole, these Value Levers comprise a comprehensive framework of factors conducive to predictions about AI project success.

For each Value Lever, we've identifi ed sub-levers–listed below–which further explain how these factors impact an AI project. We'll examine each of these Value Levers and their associated sub-levers in detail in the following chapters.

Problem

- Structured
- Recurring
- Deployable
- Interconnected

Data

- Availability
- Quality
- Cost
- Latency
- Risks-Ethics
- Risks-Privacy & Regulation

Technology

- Orchestration Technology
- Insights Platform

Talent

- Citizen AI Scientists
- Technical Experts

Execution

- Resource Availability
- Change Management & Feedback Loop
- Leadership Support
- AI Adoption Culture

Exhibit 1: AI Impact Model Value Levers and Sub-levers

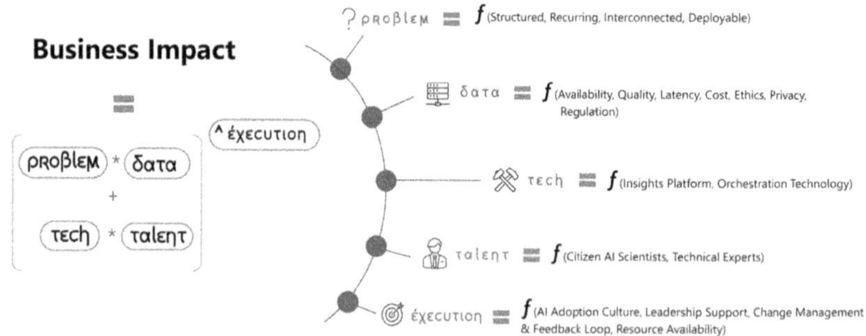

Please note that the formulas in the previous page are only illustrative and not the actual formulas that the model has found in terms of the relationships between the levers.

The model ranges from descriptive to predictive, including machine learning projects and generative AI projects. Additionally, we ensured that the respondents who helped us assess these projects comprised a diverse set of roles within their organizations, including business leaders and analytics and AI practitioners, and their input was then thoroughly reviewed by the leadership of those organizations. The methodology we developed was used to assess business impact for each one of these projects.

The assessments were conducted on a five-point scale, and we defined what each rating on the scale meant in terms of business impact. We provided a key for each rating of business impact achieved on a scale of 1 to 5. This was done in reference to the metric that the company was trying to influence with that project and against the business case established before starting the project. Similarly, for each category, we asked all respondents to assess each one of these eighteen variables for that project.

We provided a description for every point rating for every variable. For instance, what does a 5 in the data quality sub-lever mean? Or what does a 4 in data quality mean, etc.? Similarly, for the change management and feedback loop sub-lever, what does a 5, 4, 3, 2 or 1 mean? We provided a description for each rating number to enable raters to be consistent in their rating. We collected this rich data relating to business impact from hundreds of respondents across multiple clients. This included tracking the eighteen variables or subdimensions based on the state that they were in at the time the initiatives commenced.

Model Development

Exhibit 2: The Model Development Process

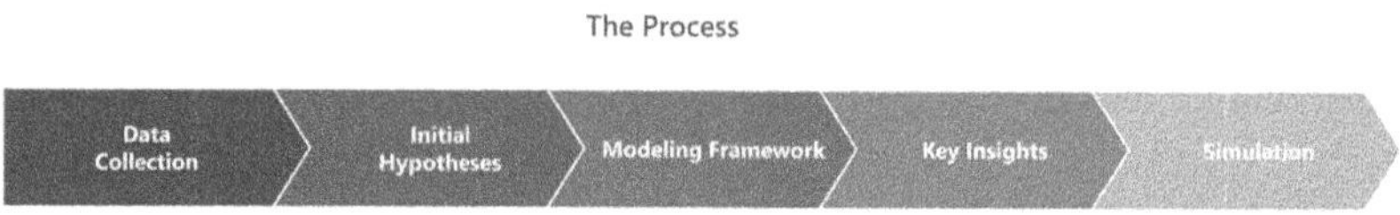

The Process

Data Collection → Initial Hypotheses → Modeling Framework → Key Insights → Simulation

Our collaboration with several enterprises was crucial to developing the AI Impact Model. These included global corporations across a variety of industries. In developing and implementing the model, we prioritized effective buy-in and support. By communicating the value of the AIIM model to business decision-makers and AI practitioners, we ensured a productive implementation process. As a means of gathering buy-in and support, we focused on explaining the benefit of delivering to our partners an actionable framework enabling them to measure and boost the business value associated with their AI projects.

Data was gathered for the model by collaborating with our enterprise partners to compile a roster of diverse AI projects encompassing a variety of business functions. To measure project features and assess business impact, we created a survey instrument. As discussed, measurement of each variable was conducted utilizing a five-point scale. Additionally, verbatims connected to the project were collected for the purpose of providing qualitative insights and context.

In developing the factors and their corresponding subfactors, we began by using our extensive experience in the industry to create an initial list. We fine-tuned this list in conversations with clients, which resulted in amendments and additions to the initial list. This included conversations with our customer advisory board, and consultation with Dr. Sawhney and major customers. The collective experience of the clients involved from their own experiences helped us refine the list.

Feedback from people at the companies involved in the testing provided us with the raw material relating to how these things go wrong including poorly defined problems, low quality data, resistance to change, lack of alignment, etc. We tried to, as far as possible, draw a straight line across those subjective opinions to make them objective by providing a rating key, where for each one of those eighteen variables, as described above, we provided a key where we very clearly defined what it means to be a 5, a 4, a 3, etc.

One question we are often asked is how did you come up with these five? Why not six? Why not seven? The answer is that we aimed to find a broad collection of factors in line with MECE criteria, which stands for mutually exclusive and collectively exhaustive. The idea behind MECE is to find variables that comprise independent and distinct dimensions, without missing anything germane to the subject. This is core to the AIIM framework's credibility, getting the dimensions right without being redundant or missing anything important. The result is a set of factors and subdimensions that all need to be considered when planning an AI project.

The factors were based not only on our own experience but also on existing literature and best practices. They were based on frameworks that have looked at what drives the success of technology projects. Thus, in terms of choosing the high-level factors, the ones making up the model are not all that surprising. In fact, it's the other way around. We wanted to make sure that we didn't miss anything; that we were grounding our definition and development of these core factors in the literature, in what people had actually done along with our own research. The goal was to cast a wide net and look at all the relevant literature. Taking these steps and performing this research led us to come up with this list, which we believe is comprehensive.

The combined list of sixty-two projects comprised the entire spectrum of business functions including sales, marketing, supply chain, enterprise technology, finance, etc. The complexity and the maturity of the projects selected also varied. Some were fairly simple in terms of descriptive analytics, which involves using data to tell what has happened: when, where, how much, etc. Others involved more diagnostics where the projects were actually getting into the root causes of problems–gauging if they were occurring, then predicting what was likely to happen. This could involve demand forecasting, sales lift or supply chain forecasting accuracy.

There were other projects that helped corporations take actions to orchestrate desired outcomes–e.g., what promotion to offer to which customer, how much inventory should be ordered to optimize for carrying cost and business needs. We also included some projects that used the latest AI developments, including generative AI, like optimizing marketing for a certain audience or a natural language system to ask business queries from organizational data.

The sixty-two projects were spread across the complexity spectrum. Who were the respondents? They were business leaders who were responsible for running the projects, ensuring that the right kind of investment was made, and also the AI practitioners who did implementation work on those projects. Generally, they were responsible for ensuring the how part or the execution part of the project.

We also learned from several of our internal projects and did a leadership review within C5i. We looked at several projects for customers other than those already included in the project. The goal was to see what kinds of projects they had done, whether they had succeeded, and the underlying reasons for the results.

The following exhibit graphically illustrates the relationship between Value Levers and project success from the data we collected.

Exhibit 3: Relating Value Levers to Project Success

Our initial analysis offered some key insights about drivers of success

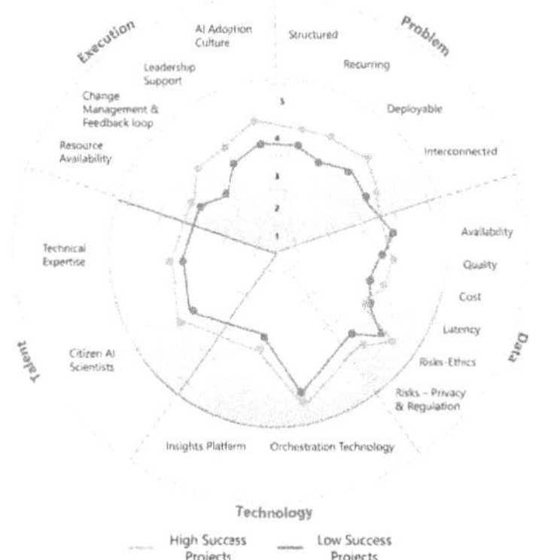

- Clarity and support **from leadership** along with a **well-defined change management** process
- **Well-structured** problem statements and **deployable use cases**
- **Data & Analytics Culture:** Democratize Decision intelligence by augmentation and automation
- **Data Quality** and **Cost**
- Presence of **Citizen Data Scientists** in the organization

The above exhibit shows some interesting analysis of our research where we took the average of each factor's rating across all high success projects and mapped it on the green line and the average of each factor's rating for all low success projects and mapped it on the red line. From this visual depiction, one can assess the impact of a factor on project success by seeing the gap between the green line and the red line for that specific factor. For example, Change Management and Feedback Loop have a large gap and that would mean that this factor has a substantial impact on success.

We followed up this simple analysis with a more complex AI-driven analysis of the factors using advanced modelling techniques. Through this analysis, we determined the statistical influence of each value level on project success. This enabled us to build a predictive and prescriptive capability into the AIIM framework.

Prediction and Prescription

The purpose of the model is to help enterprises determine when to use AI and how to ensure it meets business goals. Then, it helps answer the question: what is the likely return on investment you're going to get on an AI project? Essentially, the model predicts the business impact and allows companies to determine which use cases are most likely to benefit from implementing AI.

We'll cover the simulator used to project AI project perfor-mance in Chapter Nine. It answers the following question: if you don't like the results you're getting with this project, what levers can you push to improve your chances of success? The model doesn't just say yes or no. Instead, it might say, "The predicted success is low, but if you invest $XXX in this area, then you could enhance your predicted success by $YYY." For example, you might now have guidance on whether you should invest more in change management or data strategy, etc., in order to realize a better outcome.

"I think one aspect of AI initiatives that I've seen is there is a lack of clarity on what problem they're trying to solve. In some cases, the AIIM impact model score was less and the engagement score from the business was less. And one of the biggest reasons was lack of change management, lack of training of the team, of the associates. Those are some of the key aspects of developing AI projects that I've learned. I see there are so many AI capabilities large organizations are implementing, but can we have an objective evaluation, based on the selected factors, of what is working and what is not? That's where the AI impact model has been, true to its name, very impactful."

Deepak Jose – Global Data, Analytics and AI Executive. Vice President, Head of Data and Decision Intelligence, Niagara Bottling; formerly served in executive roles at Coca-Cola and Mars

The model and its associated simulator are both predictive and prescriptive. Prediction tells what you're likely to achieve. Prescription tells you that, if you do these things, you're likely to do better or worse. At the end of the day, that's what the majority of enterprises are most interested in–how they improve their chances of success. It's good to know whether or not this project is likely to be successful. But what's even better to know is how you can improve it, or should you even undertake it in the first place? Then, once you decide to undertake it, how do you make it work as well as possible? How do you optimize it? Or you could go to a company with an existing AI program and use the simulator to optimize it.

The AIIM is essentially using data-based evidence as a guide, along with best practices and experience, to structure the problem as to what works. The other approach is to have people rely on their intuition, their gut feel, their heuristics and their own judgment on what drives success. But it's best to ground this in real data–the facts. That way, when you put in the data to use the simulator for a specific project, you can provide all the ratings for that project, and it can predict the likely level of success. The prescriptive capability is where you tell the model: "I want to improve the success of this project, and I have so much budget available to spend." The model might prescribe, "Okay, if you have $5 million to spend, you should spend that more on data or more on execution planning or more on talent or more on technology." How you should distribute that spend across these different areas is the prescription the model provides.

AI Impact Model

We consolidated data from the sixty-two completed projects and evaluated multiple data science and machine learning approaches to understand which factors most strongly influenced outcomes. After comparative testing, a Gradient Regression (GRG) model emerged as the best-performing approach, delivering the highest statistical accuracy and stability across validation tests.

The GRG model is well-suited for complex, multi-variable environments where outcomes are driven not by single factors, but by the interaction between several variables. Rather than treating inputs independently, the model estimates coefficients capturing individual contributions of each variable and their mixed effects to arrive at a consolidated coefficient. This allows us to consider a quantified view of cause-and-effect relationships, not just correlations.

To complement this analysis, we also applied a Random Forest model. Random Forest is an ensemble machine learning technique that builds multiple decision trees on different subsets of the data and aggregates their predictions, thereby handling non-linear relationships and reducing overfitting. Random Forest was employed to validate the robustness of the GRG findings to identify whether drivers of impact were consistent across modeling techniques.

Together, these models provide a balanced and defensible impact framework combining interpretability and statistical rigor from GRG with robustness and non-linear pattern detection from Random Forest, resulting in a reliable, data-backed view of what truly drives performance across AI projects.

Why This Matters for Business Leaders

This approach moves impact assessment from opinion to evidence. Instead of relying on isolated success stories or single metrics, leaders get a clear, quantified view of which potential levers and combinations drive desired outcomes. It supports better capital allocation, sharper prioritization of initiatives, and more predictable returns from AI investments grounded in data.

In the following chapter, we'll dive into an examination of the first of the Value Levers that drive AI project success: the problem (also known as the North Star).

CHAPTER THREE

The Problem: The North Star

"AI is only as powerful as the problem it is pointed at, if you don't define the why well, and if you don't frame the right whys, you will never get to the right outcome. AI doesn't fail because of the model. In my mind, AI fails because the mission doesn't define the human why driving the project–once you define the human why, technology will find its purpose."

**Ram Iyer – General Manager,
Windows Devices & Sales, Marketing, Microsoft**

The North Star of any AI project is the problem it is designed to solve. Our research has identified properly diagnosing the problem–the first Value Lever–as one of the main factors responsible for the success or failure of such projects. Achieving clarity and alignment around the problem you are solving is vital to generating business impact from AI. If you don't define your problem precisely when engaging in any kind of innovation initiative or AI project, it's like the line from Alice in Wonderland: *if you don't know your destination, any road will take you there.*

Another adage that applies here is the saying that *a problem well defined is a problem half solved*. Understanding the use case and accurately defining and scoping the problem is crucial to successful AI implementation. A major reason for project underperformance is that people get impatient and jump into implementation. In most cases, they would have been better served to wait and spend more time upfront defining the problem. That is a clear message revealed by our research.

Within the larger "problem" Value Lever, we've identified four sub-levers that influence whether the problem an enterprise is trying to solve is likely to be successfully addressed with AI. In this chapter, we'll take a deep dive into each of these subdimensions in order to explore how addressing them drives AI project success.

These sub-levers are:

- **Structured**: The extent to which a problem has clear parameters, a defined scope, constraints, and success measurement, allowing AI systems to generate consistent and actionable outputs.
- **Recurring**: Problems that recur with greater frequency are generally more amenable to being solved through AI solutions. When solving problems with lower frequency, there are often other nuances that need to be considered.
- **Deployable**: Measures the scalability of the solution to the problem and how easily it can be integrated into existing workflows. Projects scoring highly in this category have been found to drive greater business impact.
- **Interconnected**: The degree to which the problem is connected to other problems and systems. Such connections enable the solution to touch various aspects of the business; thus, solving it can create valuable side effects for other problems.

Sorting the Sub-levers

At the high level is the problem, but as outlined above, within this Value Lever, there are four sub-levers or subdimensions. We measured the impact of each one of these and used the results as a variable to predict success. We knew upfront that if the problem was well structured, implementing a plan to address it would likely be more successful. However, we've let the model guide us and haven't pre-judged any particular outcome.

The eventual model built on historical data of sixty-two projects evaluated how important a particular factor was. The results made it clear that having a well-structured problem is a critical variable in driving success. This was an important validation of what we already surmised from our experience. The model tells us that interconnections are important, but not as important as the recurring nature of a problem. Recurring problems are more likely to be successful, and a solution to a problem that is well structured and deployable is more likely to be successful. So, among these four: recurring, interconnections, structured, and deployability, interconnections is an important category, but not quite as important as the other three.

In the following sections, we take a closer look at each of the subdimensions of the **Problem** Value Lever.

Structured

We have identified several areas that need to be addressed in the process of determining how well structured a problem is. To structure a problem, first of all, there needs to be an objective. Ask: what is the objective this problem is solving for? Next, we need to define the problem that we're solving for. We also need to ask what its scope is, which relates to the breadth and depth of the targeted business impact. Which markets? Which products, etc.? Then, constraints come in, so we ask: what are the constraints, the guardrails, the boundary conditions? How can we define them?

Typically, if we define an objective and do not define scope, constraints, and structure, the effectiveness of the project suffers. If none of these are defined well, we have identified them as deserving a very low rating of 1. If the objective is defined, but others are not defined, the rating is a 2. If the objective and scope are defined, but constraints and structure are not defined, it is likely a 3.

Typically, that's the sequence in which you will work on a problem. Define an objective and then define the scope. So, a high rating would mean that the objective, scope, and constraints are defined, but the structure is not clear. A 5 rating would mean everything is well defined, and the problem is very well structured.

When you structure a problem, it's vital to define its scope. This includes identifying and defining the constraints, the problems you're solving for, the things you're not solving for, the things that you can change, and the things you cannot change. That's how a problem is structured. For instance, if you want to increase sales, that's a very broad statement. But if you say, "Which are our best-selling products, the ones that we should focus on?" or "Which products should we offer discounts on to increase our sales?" Those might be better ways to define it. The takeaway here is that it's crucial to define the problem as specifically as possible.

While normally it is important to structure the problem well by defining it specifically, there is one exception. This happens when a POC (proof of concept) is necessary. Sometimes, there are projects that are exploratory because you are interested in examining or investigating a hypothesis. But typically, you want to do projects of this type on a small scale. Use them to help you nail down and better structure a problem and eventually create a bigger initiative on a tightly defined problem. A Silicon Valley term for this type of approach is *Minimum Viable Product*.

In terms of structuring a problem, one situation where this becomes difficult occurs when you're simply dealing with an unknown set of issues. You're trying to do something very new, very different, and you don't want to tightly define everything because you may not be defining it correctly. If you try to define it very specifically, you may miss out on something. This can occur when it's a fairly unknown industry or unknown space for the enterprise involved, or there's a new product line, etc.

In those cases, one thing you can do is to launch a few POCs. Structure each one around a different approach and try to use those POCs to narrow down how to more clearly structure the problem. For example, if a big issue for a company is, "How do we make sure that our Brazil launch is successful?", it could be, "What product line should we use?", "What should the marketing message be?" or "What pricing should we use?" There are many different things that influence that launch, but you may want to narrow it down and say, "Okay, these things we're going to keep constant, we want to charge this price, and this is what our brand says. But we're willing to tinker around with our product design." You can't have AI trying to predict everything altogether. So, launching a few POCs can help you better structure the problem.

There is something to be said for situations that are unstructured because sometimes structuring a problem too tightly, without doing enough experimentation, can lead to structuring things the wrong way. It's not that structuring a problem well is not important, just that in some cases, there isn't enough information available yet to structure it as specifically as you would like. Thus, POCs can be helpful when you're dealing with many unknowns. For instance, if there is fluctuation in your business due to a high degree of seasonality, that generally shouldn't be an issue. But if the industry is going through significant change, simulation can become difficult. And given that AI does prediction and simulation, it then becomes more important to tightly define the problem.

The final decision will have to be made by you, the manager. The AI may be able to give you something like five or so different versions to consider. For instance, "If some factors remain constant and others change, then this will be the likely outcome." Or "If other factors change and these remain constant, these different scenarios will be created." This approach enables you to structure a variety of factors, create a number of scenarios, get multiple answers, and make the final decision yourself.

Recurring

A recurring problem is simple to identify. It's not a one-off issue. Then, we ask how often it recurs. Are you employing AI to inform a certain strategic decision that a CEO or senior management is going to make once and only once? For instance, "Should we go into Brazil as a company or not?" Or is it something along the lines of, "The problem is that every time a customer comes to buy this certain product, should it be priced differently for that particular type of customer? And what price should we give to which customer segment, or what discount should be offered?" That could happen every day, even many times a day.

Problems that occur repeatedly are prime targets for an AI solution. Thus, it's important to understand how often the problem occurs. On a scale of 1 to 5, a lower number would mean less often and a higher number more often. For example, if you are going to refresh your product portfolio and the problem centers on making changes to the product portfolio, the recurrence of the problem, depending on the industry, might be once a year. Whereas if the problem you're trying to solve for is assortment planning, you might need to deal with it on a biannual basis–twice per year.

As we move to very high recurrence, this might involve something like an IBP (integrated business planning) process where every month marketing, finance and production meet and discuss the market demand and associated resource allocations for the company's products or services. In this type of meeting, production might say, "This is what we can do. This is our capacity, and these are our constraints." Generative AI platforms can also be used for such problems to give qualitative insights from analyzing publicly available information.

However, it is important to do enough "fine-tuning" to give context to the model and keep enough of a "human in the loop" presence to augment the findings and check for mistakes and hallucinations.

While the recurring nature of a problem has an influence on success, as with other factors, there can be exceptions. Generally, recurring problems do have better solutions because you build the model once, you perfect it, you fine-tune it, and that enables you to keep solving the same problem. However, there may be projects that recur less frequently; sometimes, those may be very important initiatives which hold substantial value and also lead to substantial decisions. In those cases, leveraging AI can be important.

When programs that are one-time or have low recurrence are initiated, the emphasis on some of the other factors of the AIIM may need to be stronger to make sure that you're getting sufficient value from the program. Defining the problem well is crucial if it's a one-time problem or something happening once a year because you only have one opportunity to solve it. In this case, you really need to structure it well to maximize the project's chances of success. As we've discussed, if you tell the AI something like, "Find out how to make sure I hit my revenue goal this year," that's much too broad a request. You need to define very specifically what you want the AI to help you predict and optimize.

The takeaway is that we are not suggesting enterprises should never invest in non-recurring problems. These can concern very important issues that need to be solved. But when dealing with them, it's important to make sure to work on making the other factors stronger.

Deployable

Next comes deployability, which considers how well an AI solution can be deployed in the system of engagement through linkages to the system of insights in a production environment. It centers on asking the following question: Can you actually take this solution and deploy it in a production environment where it is easy to use, relevant to the business and can solve the problem? If deployability is very low, neither of these–accessible in a production environment and linked to the system of action which can address the problem–is the case. Very high means that both are the case.

The system of insights might tell you, for example, what offer or message will enable you to sell a certain product for a certain microsegment that you have created. This is an insight. Action is telling the person who is supposed to go and do that to get it done. For instance, the action would be to tell a person in a certain geography that, "In your area during this period, you need to run these promotions for this product." If you don't have this type of linkage, it means there's no person to tell or there's no way to convert it to action. It would require much more manual effort and be subject to a significant amount of confusion as well as potential missteps. Thus, the insight is not properly acted upon and, therefore, the result is not what you would expect.

Enterprises usually have three types of information technology systems. Generally, there is essentially a system of records that serves as the foundation–typically an ERP or CRM system that records every transaction that happens in the enterprise or in a specific department. On top of this, there is typically a system of engagement that manages interaction with the different entities within an enterprise. Related to both is a system of insights that suggests intelligent decisions and actions based on the data it receives from the system of records and the system of engagement. What we have found is that the initiatives that are seamlessly linked across these different enterprise systems from the data to the AI models to the execution are more likely to be successful.

In other words, you have to traverse up and down your tech stack. Only then will you be able to achieve implementation and deployment at scale. The reason is that if you are not connected to the system of engagement, you might find an interesting insight from the AI system but be unable to transmit it into an automated workflow. That's where the action takes place. And whether this system of engagement is based on AI agents that can automate tasks and perform them, or traditional machine learning AI, if you can't bring that insight to fruition in the form of execution, you won't be able to realize the true value of AI.

Exhibit 4: Integrating Systems

Integrating systems of insights with systems of engagement/records

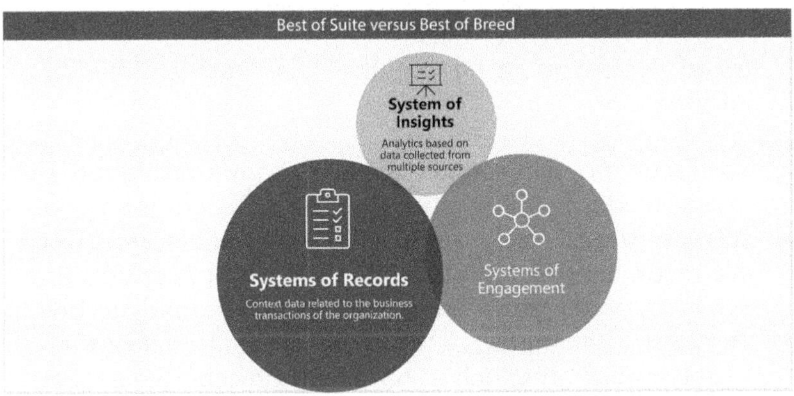

This is where the issue of choosing between best-of-breed (the best tool of its type) and best-of-suite (the best suite of integrated tools, considered as a whole) arises with respect to deployability. Deployability centers on the ability to connect different types of systems within an organization. The algorithm is then able to run through those systems and input the data into the model to build it, and the model orchestrates the outcome; then, that outcome generates feedback, which again goes back into the model. In that entire process, there may be some situations where you can't complete the loop because your model is only giving an insight to a business decision-maker who will then make the decision.

The question is: are you at least able to get that person the right information at the right time? Are you able to capture the decision that person took, track the outcome of that decision and create a feedback loop to the model? The takeaway is that the greater the level of deployability, the more helpful the proposed solution to the problem is likely to be, especially as it relates to the ability to tie together different systems.

Scenario: Integrating Deployment at Scale

Industry: Consumer Packaged Goods (CPG)

Objective: Make real-time data available to reps in the field

Issue: System must tie together disparate sources of information

This scenario focuses on an AI program designed to integrate deployment at scale. The CPG company has salespeople on the ground selling to retailers, reps who are responsible for ensuring that the right products are available on the shelf when consumers are ready to buy. They carry technologies and systems that enable them to accomplish this such as handheld devices like an iPad or a smart phone.

Making this happen hinges on deployment via the system of engagement. The company's sales reps would go to Walmart and other stores to ensure that the product was available for purchase. This wasn't an automated process, so they actually had to look at the aisles in person. Their old model devices would scan the shelves and provide them with some degree of stocking information. But the newer devices the company introduced, with AI built into them, could not only tell them which products were running low but also give them the ability to place an order right then and there if needed.

The actionable insights they might receive would be that some products are much more likely to be successful and run out of stock than other products. So, it's key to make sure that you keep a buffer stock of those items in your back room or in a place where you're staging your inventory closest to the retailer. Thus, when you order, those products are either immediately available or take the least amount of time to get there. As a result, you avoid losing sales due to a lack of availability.

Modern technology can use cameras placed in the aisles pointed at the shelf to determine product stocking levels. Then, algorithms called AGA can be used to measure the height of the shelf where the products are placed. If that height goes below a certain level, it notifies you that it needs to be restocked. A similar technology allows sales reps to go through the aisles using their own device to scan the shelves, which triggers a replenishment signal in that device in the order-taking system. So, different levels of technology are available to perform these tasks. In this case, the handheld device doesn't just ingest data, it is also connected to a system for placing orders. Another type of device captures images, and those images will trigger replenishment of stock if needed.

This company used devices that enabled sales reps to gauge stock levels and place orders to replenish them when called for. The system would then suggest orders based on this data but would not place them automatically. That allowed for human judgement as to whether placing the order in fact made sense. The sales rep could utilize data pertaining to where the orders would be shipped from, for instance, they could ask which of our warehouses that have those twenty bottles we need will be cheapest. Additionally, the rep could ask questions such as: How long is it going to take to get there?

Do we need to rush? If you need it in a hurry, that's one thing, but rushing costs more. If you don't need to rush, it's better to take it from further away or use a slower UPS route to save money.

Thus, the technology provides the sales force with insights and guides them on what actions to take and then enables them to take the actions and ensures that the products are there. That's an example of deployability at scale. The sales reps are looking at which products are not moving that fast, which products are going really fast, and which products they should prioritize in restocking. This is an example of systems of engagement and insights and records coming together.

The lesson here is that combining your system of engagement with your system of insights and system of records can significantly boost efficiency if properly integrated and deployed at scale. It can enable you to get things done in a faster and more impactful way.

Interconnected

Number four is interconnections. Typically, most organizations have some divisions that function as silos and often don't take sufficient advantage of overlaps or synergies. Given the way reward and reporting systems work in companies and also given the volume of business questions faced in any one department, people often do not think beyond their own silo or function. This is suboptimal because the real world is not like that. If you're going to create a forecast for a product, for example, you'll be more accurate if you know the product's pricing and promotion, not only for your company, but also for your competitors. You also want to know how to answer questions such as: Should you take the competitors' figures into account? What's the inventory that is available for that product?

Are there going to be stock outs? Are there going to be some constraints? What's the cannibalization of that product?

There are multiple variables from different functions that will impact your forecast. And if you do not connect those to answer that question, the forecast is likely to fall short. If you are not able to produce a forecast for a product during a particular period that considers price promotion, inventory, and aging as well as potential cannibalization coming from different functions and different systems, your forecast isn't going to be accurate. To the extent you can accomplish this, the accuracy of your forecast is likely to improve.

That's an example of the interconnections of a problem and typically projects that are not very successful will be at the low end of the spectrum. So, very low to low, a 1 or 2, means not connected to any other function or perhaps connected to just one function. If you are at 3 or 4, you're generally doing well in this regard, with two or more connections, while 5 is excellent, with a maximum of possible connections.

If you're trying to solve a promotions problem, in the sense that you're trying to drive revenue from your retail partners, and you're deciding how much in the way of promotions to offer and to whom, that also has something to do with your supply chain and the availability of stock. If you offer too many promotions and then you can't meet the demand, that won't help either. So, the issue of promoting your products is connected to another problem and another part of your organization. This is what we mean by interconnected problems as opposed to discrete problems that are limited to a single department or area of your organization.

A real-world example could be a company that wants to use AI to improve its cash flow. For instance, if a company in a services business in the consulting world needs to project cash flow, they have to look at not just what's happening with respect to their capabilities, but also the market conditions in the field they operate in. Are they expected to be good or bad? Is there going to be enough demand? Do their employees have the skills to provide the appropriate services to the market? How much support do they need, and do they have the right capabilities, technology and support staff to successfully provide consulting services? To determine this, they need to look at multiple variables and combine that analysis to generate a cash flow forecast.

When you solve a problem of that type, if you're only connected to one function and you only look at what the market is expected to do without also looking at the skill sets of your employees, for example, you are unlikely to get a complete picture. In that scenario, if the company forecasts its cash flow based just on market conditions, it's not likely to be accurate because they have not considered various other factors involved in accurately projecting cash flow. When there are more connections, it's a more comprehensive solution. To bring data from all of these functions, collate it, create some sort of abstraction from it, and finally realize some sort of insights out of it is a very complex process. But if the company does it properly, their forecasting is likely to be much more accurate, whether it is a demand forecast for their services or a cash flow forecast.

Delivering insights from this type of complex analysis is a strong point of AI. As little as ten years prior to the writing of this book, it was technologically challenging to analyze data in this way. This made it very difficult to bring all of the data to one place and then have the technology on top of it to generate insights or a solution from that data. Today, the technology to accomplish that is here, making it feasible to implement highly interconnected analysis. Generative AI models can be used to research related environmental factors which can be combined with internal data to provide a consolidated insight or point of view.

In analyzing the interconnection factor, the conclusion generated by the model is that the more systems and departments the problem touches are recognized in the planning process, the more likely it is that you'll generate positive ROI from AI. It works both ways, in that it's more likely to drive impact because it can affect various parts of your business, and there could be downstream benefits that might lead to additional projects elsewhere. At the same time, it could actually be more likely to fail because of the fact that it requires cross-departmental collaboration. Different departments mean different types of data sources coming in. Each data source has its own structure and its own channels, and sometimes the data sources don't effectively interact with each other.

Those different departments, in terms of your organizational culture, have to cooperate and collaborate. So, if you're making a decision on what promotions to run for your product–to, for instance, sell off your inventory–it has to materialize as a combination of your marketing department, your merchandising department, and your supply chain department. You can drive greater success if you tie those things together, but if your departments have different data structures, different systems, different processes, different cultures, and different people who aren't used to collaborating, it can be very difficult to do. Again, interconnectedness often, in some way, feeds back to culture, so all these variables also have cross-linkages, and some of them work in combination with each other.

Case Study: Defining the Problem Well Leads to Productive Insights

Company: Global technology company

Objective: Fast turnaround and productivity in delivering insights to management

Issue: The problem needed to be identified and structured clearly in terms of stakeholders and expectations

This company found that producing sales decks for internal and external presentations was consuming significant staff time. The decks served an important purpose, so rather than discontinuing their production, the company searched for a way to do so more efficiently. Designing a sales or review deck required substantial manual effort and resources. The decks typically followed standardized templates and contained recurring patterns of information and trends. This created the opportunity to automate the creation of these insights decks using code, thereby significantly reducing the amount of staff time spent creating them. Factors supporting the chance of success in accomplishing this goal included that this was a recurring problem featuring high data quality and was supported by management, which realized that automating this process could free up staff time that could be spent on higher-value activities. Moreover, it could generate faster output to benefit management in their decision-making.

Once the project was approved and the code had been written to automate the creation of these decks, the company quickly began to see benefits from it. What once took many hours of human labor can now be done in a fraction of that time. Staff responsible for creating the decks simply needed to enter the basic information in a standardized format and the AI handled the rest of the job. As predicted by the AIIM, this project delivered a substantial business impact and therefore a sizable ROI when measured against the time and effort needed to create the code that made the project possible.

The takeaway from this case study is that this AI project's success was predicated on an accurate analysis of the problem and the expectations from the AI initiative. This was a recurring problem that, if it could be solved, presented the opportunity for saving significant time, and therefore money. Thus, defining the problem well served as an effective first step towards solving it. When that problem definition success was combined with the necessary resources and leadership support, all the pieces of the puzzle were in place to secure a positive business impact from the project.

Avoiding the Temptation to Jump to the Solution

A key mistake many organizations make is not adequately defining the problem before launching an AI project. The failure to spend enough time evaluating the problem often occurs because of an eagerness to jump to the solution. This is a common mistake, the tendency to propose solutions even before we understand the problem. An analogy for this comes from product management, where it can be said of practitioners that, "You're not a product manager, you're a problem manager." This saying encapsulates the reality that, to be successful as a project manager, what you have to stay focused on is the customer problem. Your product is just one way to solve the problem. Don't jump to the solution before you've understood the problem.

It's very tempting to jump to the solution. For example, if the IT people or the AI data science team are in charge, they'll often jump to building models and deploying technology mapping solutions. However, the most important question is, what is the problem to be solved? Determining that requires a solid understanding of the context, the business domain, and the business use case; it also requires patience, which can be a challenge because it can be seductive and easy to jump into solutioning. This is where it's important to realize that if you want to move fast, you first have to slow down. Which means, in practice, that organizations should look to properly define the problem first. Thus, a key tip for successfully implementing AI–or any technology–is not to rush to the solution.

Get the people who have problem information and problem expertise in the room and on the AI implementation team. For instance, if you're building an AI project that involves helping your salespeople, get them in the room because they can provide valuable insight into the problem. It's crucial to understand that the problem expertise sits close to where the problem occurs and not where the solution is being built. It's vital to bring the people with that understanding into the conversation early and continuously. Bringing in everyone with relevant knowledge helps generate buy-in for the project as well as enabling the gathering of relevant information. The goal is to ensure that you're importing enough problem expertise and problem insight to make the project a success.

Putting the *Why* First

To drive success in any business initiative, the *why* needs to come first. And the: "Why are we doing this?" or "Why are we doing that?" in a business technology context needs to highlight that technology is in service of a cause–and the cause is the business. The business has goals, the business needs to grow, the business needs to acquire customers, the business needs to rationalize operations, etc. These are the business goals and strategy, and from there the business problem arises, and the business problem is the *why*. The technology is the context within which the *why* can be addressed and the problem solved. The *why* should never be to use technology for technology's sake. The *raison d'etre*, technology's reason for being, is to solve a business problem.

Therefore, we start with defining the problem and thus this chapter comes first in our discussion of the five Value Levers. Fortunately, we also find that the importance of defining the problem is empirically borne out by the model. To put it bluntly, if you do not define the problem correctly, if you do not know what the North Star is, everything else is likely to fail. In this regard, we might say that a headless chicken moves with tremendous speed–but with no purpose. This analogy serves as an example of how not to implement AI programs. As alluded to earlier, we see this sometimes in organizations where the technology division, the CIO and the IT department, drive the process. When they lead AI initiatives, what can happen is that the solutions get ahead of the problem. The cart gets ahead of the horse, and you can end up with solutions looking for a problem.

For example, the optimal approach to AI is not to say something like, "Listen, we are going to create an AI system that will help HR." Instead, it's better to start by saying, "Listen, interviewing is a problem for us. It's costly. We're making mistakes, we're hiring the wrong kind of people, and it's very expensive. How do we solve this problem?" That's likely to be a better approach–to let the problem drive the solution rather than vice versa.

Another factor that should inform AI project design is the framework of jobs to be done. This gets to the heart of what function AI will serve in an organization. The CEO of Cemex, the cement company, has said something to the effect of, "Nobody wakes up in the morning and says, 'I want to buy a bag of cement.' Instead, they say that they want to build a whole road or a house or a cathedral, and our job is to help them do that." That gets to this idea of: what is the job to be done by AI? What is the problem to be solved? That is the *why* we need to start with. That's the core idea, the North Star. Everything else falls under that.

Generative AI and Structuring the Problem

It is possible to use some of the GenAI platforms available at the time of the writing of this book to get support in structuring a problem better by providing sufficient context to the platform. However, these are still developing and should be used for such an application to augment humans and not to replace them. This may, of course, change in the future as these platforms continue to improve.

For instance, one way of using generative AI in the problem definition stage would be to gain a better understanding of the issues involved to make sure you aren't missing certain aspects of the problem. You could also use GenAI to write the problem in terms designed for different audiences and stakeholders; you might need a technical version for developers, then, you might need a business version or a customer-facing version. So, you can certainly use it as a thought partner. What's essential to the process of getting accurate and precise context is that you have to provide it with proprietary data. You have to provide it with your own corporate information because otherwise, the model is only going to know about what it has been trained on.

This involves adding an element of context engineering by providing it with relevant documents. In terms of context, what you can also say is, "I want you to read all these research reports, all these documents, internal documents, and then use that information to better structure the problem." That's called accessing a knowledge base rather than simply providing it with the knowledge.

Problem Definition and Agentic AI

Problem definition and problem selection are crucial to achieving success in an AI implementation. If you're using an agentic AI approach, this takes on a very different flavor. The reason being in agentic AI you have to decide, firstly, to what extent you're going to use agents and what combination of agents and humans you plan to use in the process. Are you going to have more humans in the loop, or fewer humans in the loop? Secondly, the selection of the problem is key. Are you selecting a problem that has high risks attached to it? If things go wrong in the medical field, for instance, the results can be catastrophic. Or even within your business, if you are selecting something such as using agentic AI where you're going to create strategic communications to send to certain of your key clients in a B2B business, it can backfire if the AI is not properly trained.

For instance, in a business where you have clients who are each worth multi-million dollars, and you're selecting communications to send to key stakeholders, something going wrong would have huge implications. On the other hand, if you have millions of prospective customers and if you're sending communications to all of them, and you don't send a perfect communication, as long as you don't send anything offensive or that will significantly harm your brand, it could still work. It might be fine as long as, say, ninety-five percent of the recipients think it's good. It depends on the situation, and the tolerance for risk.

This is because the nature of agentic AI enables you to actually drive interconnected problems more effectively. Because you can use different tech stacks, and you can have multiple agents involved, problem selection and problem definition become even more important with agentic AI approaches.

At the time of this writing, there is significant industry wide development of agentic AI approaches and some early-stage deployment at enterprises. With increasing use of agentic AI, we anticipate that in the near future it will become easier to create linkages between disparate systems within the enterprise and organizations that adopt these approaches will score much higher on deployability.

Earlier in the chapter, we identified deployability as a key factor in the model. It's vital because you're tying together the system of engagement with the system of records and the system of insights/knowledge. Generally, organizations have multiple platforms and systems they have subscribed to and tying them together can be a mess. They've got one system for insights, they have another system for records, they have another system for orchestration and launching campaigns; at some points they may be using SAP, Adobe, Salesforce, etc. or some combination of these.

The beauty of agentic AI is that it can be used very effectively to tie together multiple systems because the agents themselves can extract information from one system and export it to another. They can talk to multiple systems faster than a human can. Deployability has clearly come out as an important driver of success in our model. Perhaps, with agentic approaches, deployability may not remain as much of a challenge a few years down the line because agents will enable greater deployability of AI models. That's one thesis.

Scoping the Problem and AI as a Tool

After we identify a problem, we then have to define it. We have to scope it. We have to put boundaries around it, because you can't solve every problem. *You can't boil the ocean*, as the saying goes, so framing the problem is the next issue. Identifying the problem is important, but so is framing it which involves scoping it, further defining it and putting boundaries around it such as "This is what we will do, this is what we will not do." Or "This is a problem we will solve, while these are the problems we will not look at." This can involve narrowing the problem or sharpening your focus on the core aspects of the problem you want to solve. If a business has multiple use cases that need to be solved with AI, and they are interrelated, it is better to take an incremental approach rather than trying to go "big bang" and solve everything in one go.

It's as if a problem is a diamond in the rough, and you need to turn it into a diamond. Alignment in addressing the problem is where everybody, even if they don't agree, buys into the solution. It's similar to the principle from Amazon, "Disagree and commit." This means that you may not all agree, but you have to commit. Everybody needs to be aligned with what you're doing.

One organizational issue that can lead to poor problem definition stemming from corporate structure and interdepartmental communication is a lack of connectivity between the silos. To illustrate this, we can go back to the description of how the domain expertise and problem insight and expertise sit with the salespeople in the field, but the idea of what solutions can be built sits with your AI teams and IT departments. You have to bring these people together to optimize your ability to implement an AI project successfully. If that doesn't happen in practice, we can look back to the parable of the six blind men and the elephant. You don't connect the dots, and if you don't do that, you're never going to get to the answer.

The silo problem prevents you from breaking through to come to a holistic, comprehensive solution. The solution to the problem you are solving for benefits from taking a deep dive into how you define the problem and how the nature of the problem affects the impact of that AI program. The practical guidance around this issue centers on identifying the problem and specifying its scope. It comes down to identifying the right problem and defining it properly.

A simple example is that, sometimes, when launching AI initiatives, organizations don't clearly define what they're trying to achieve. Second is that, even if you define what you're trying to achieve, you need to, within that, define certain constraints and certain levers. The better you do this, the more effective you're going to be. For example, if I have an AI problem where my goal is to increase profit, while that's a worthwhile objective, it is quite broad. It's not very specific. It's similar in this way to the earlier example of telling AI that your goal is to increase revenue.

You can increase profit by increasing revenue, you can increase profit by reducing costs, and you can increase profit by changing the mix of revenue and selling more profitable products. And then there's the short term versus long term. In the short term, I can increase profit by cutting all my sales and marketing expenditure, but that may impact my long-term profit because my organization's brand value is likely to fall. The takeaway here is that there are various dimensions to the process. This is a very high-level example, but it gets at the insight that you are more likely to be effective with implementing AI if you define your problem as granularly and as tightly as possible.

It starts with knowing what you're trying to achieve. Then, defining the key constraints within that. There are some things you don't want to touch and some things you do want to touch. Going back to trying to solve the problem of increasing profit, it is such a broad statement that it is prone to being interpreted differently by different people. There may be some things you just can't implement. For instance, if you tell the AI to cut costs and the AI comes back and says, "Fire 500 people," that may not be a feasible solution. Once again, it's vital to define your problem as specifically and granularly as possible.

Additionally, it's essential to realize that while AI may be more efficient and autonomous than other workplace tools, at the end of the day it is still a tool. One of the authors of this book, Dr. Sawhney, has been known to say, "A fool with a tool is still a fool." And with AI, perhaps a more dangerous one. Because now, you might say that the person is an augmented fool or an accelerated fool. The point is that any tool, however powerful it is, needs to be directed. It brings to mind a quote that is attributed to Einstein: *Intellect is a great horse to ride.*

Artificial intelligence is a great horse to ride somewhere, and that somewhere is the guidance, the strategy, the use cases which amount to using this tool in the appropriate way. In that context, we can say a fool with a tool is still a fool if people are not letting the problem guide the tool's application. It's similar to the saying, *if all you have in your hand is a hammer, the whole world looks like a nail*. The takeaway is that you cannot start with the tool first. You have to start with the use case, or problem. Then you find the right tool (or AI application, in this context) for the job.

CHAPTER FOUR

Data: The Foundation of AI Success

"Across the value chain of an AI model, I believe there are several critical factors to focus on. The starting point is the data itself. Its cleanliness, impact, quality, and completeness are essential elements. However, I think a lot of organizations tend to overlook the importance of governance. From an HR standpoint, we are actively driving the whole governance of AI, not just from a data quality and data completeness standpoint, but also thinking about what kind of data is actually needed and what level of information are we providing? Ensuring that no personally identifiable information is exposed or transferred. While strengthening governance can slow the process down, it's a necessary step. Ultimately, if you have a robust governance process in place it creates trust in the whole process."

Ujjwal Sehgal — Global Head of People Analytics, Mars

In this chapter, we'll shine a light on data–the fuel that powers AI. Data is the second Value Lever and plays a vital part in determining whether an AI project will succeed. We'll explore the various aspects of data that must be taken into consideration when designing an AI project, and how each plays a role in project success. For instance, there is a fine line between the need for data availability and the diminishing returns that can occur after a certain point, so it's crucial to evaluate the data a project needs from the standpoint of how much is enough and how much is too much. Quality, on the other hand, is absolutely essential to project success. Bad data can derail a project all by itself. In this context, a favorite quote of one of the authors, Ashwin Mittal, is: "In God we trust; everyone else must bring data."

Featured in this chapter is a scenario centered on the banking industry that highlights the risks of relying on flawed data and the lessons learned from the experience. Additionally, we'll discuss how readers can assess and improve their data readiness and cover the importance of adhering to ethical and regulatory guidelines when gathering and using data.

We've identified six sub-factors, or subdimensions, that are key to effectively incorporating data in an AI project:

- **Availability**: Data should be accessible in sufficient volume and variety to effectively train and sustain AI models.
- **Quality:** To ensure reliable AI outputs, the data must be accurate, consistent, and relevant.
- **Cost:** The expense of acquiring, processing, and maintaining data must be justified by the potential ROI from the AI solution.
- **Latency:** Data must be available in time or with minimal delays, depending on the application's requirements for timely insights.
- **Risks-Ethics:** Ethical concerns, such as biases or unintended consequences in data usage, must be addressed to avoid reputational and legal risks.
- **Risks-Privacy and Regulation:** Data handling must comply with privacy laws and regulations, protecting user data from misuse and ensuring compliance with frameworks like GDPR or CCPA.

Generally, each element of good data management we've identified must be present to at least some degree for an AI project to succeed. We discuss these subdimensions of the Data Value Lever in detail below.

Availability

Data availability is an important condition for AI project success. It is a fairly straightforward criterion. Is the data available? Can it be easily accessed? When we say available, we mean that the data must be accessible when it is needed at the required frequency. This doesn't mean that data is available somewhere in theory, but that it is actually available at the right time.

High availability is defined as being available for greater than the desired frequency and the period required for future needs, as additional initiatives are explored. Essentially, this means it is available at the required time, and it will be available going forward.

Our model shows that there is diminishing marginal utility of data availability beyond a certain point. Sometimes, organizations focus on the objective: "Let's just go ahead and collect all the possible data," the idea being that they will continue to amass more and more data for the project before getting started. But we've found that you need the right data–data specifically relevant to the problem you are solving for–and not every piece of data you can get your hands on. There's no point in collecting data that won't be relevant to the business question. You need a sufficient amount of the right data. Moreover, AI can now be used to generate synthetic data when real-world data is limited. An example of a use case where synthetic data is being used is for gauging consumer response to a new product concept. Synthetic data is being combined here with primary research data and social media.

Small data sets can often give you better analysis because they are targeted at the item you are testing (but be careful to ensure that your testing method scales effectively if using large numbers of records in production). Removing the outliers or any noise creates this smaller, more relevant data set.

If you have too much data, it provides diminishing returns; interestingly, more data beyond the point of diminishing returns has a negative effect on system performance. This brings up the question: why should more data have a negative correlation with business impact? The reason is that there's an interconnection with other factors. When there is excessive data, it can cause greater latency. If you keep pushing for more and more data, the new data sources needed to collect that additional data may delay availability. This also increases the cost of collecting, storing, and processing that data.

The takeaway is that data availability is important but has a diminishing returns curve and then at some point actually has a negative effect. That point can be termed data overload, where excessive data collection is weighing the project down.

Exhibit 5: Data Availability and Diminishing Returns

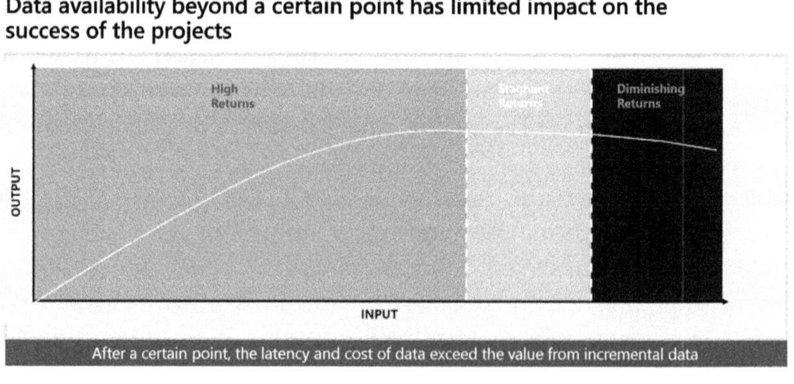

Quality

Data quality is essential, as demonstrated by the famous saying in technology circles that if you put garbage in, you will get garbage out.

To evaluate quality, we've broken this criterion down into the following components:

- *Accuracy*: The data matches calculated or supplied values.

- *Consistency*: Data arrives in the right form every single time.

- *Relevance*: Is it relevant to the task both in terms of subject matter and meeting the required level of abstraction or granularity? For instance, SKU data isn't relevant when product category data is required and vice versa.

- *Integrity*: Is the data true in the form in which it is delivered? Is it integrated to the degree necessary? Are all data elements required to calculate the data taken into account?

- *Unique*: This is a straightforward category. Is the data unique and not replicating other data sources?

- *Understandable*: The data must be able to be understood in the form in which it is delivered.

These are the six variables that determine quality levels, and we define them as very low to very high in quality. We have given greater importance to data accuracy as that is a base requirement for data quality. As long as the data is accurate, the score for data quality will be determined based on the number of the remaining five data quality criteria that are satisfied.

Exhibit 6: The Importance of Data Quality

Cost

This subcategory measures the cost of acquiring, storing and processing the data within a reasonable range. The rule of thumb we've used in scoring cost is that it should be approximately ten percent of the overall project cost. So, a score of 1 would be a very high cost of data, more than ten percent. Two is between seven to ten percent, 3 is five percent, 4 is two to five percent, and 5 is less than two percent.

In our experience, the likelihood is that data costs will fall somewhere in the middle or perhaps on the higher end of the spectrum. The expense of acquiring, processing, and maintaining data must be justified by the potential ROI from the AI solution.

Latency

Potential concerns about data latency exist across the value chain of the data handshake. Latency measures whether the data is available when it is required. High data latency means you're not getting the data when it is required. For instance, if an enterprise plans to implement production schedules to execute certain processes, but input data as to inventories and crew availability does not arrive in time, this high degree of latency will prevent the execution of those plans in a timely manner. Orders will be delayed as a result, and customers will not be happy.

Latency is not always bad or crippling for a project. When the data doesn't need to be real-time, a higher level of latency can be acceptable. This correlates to, for example, how recurring a problem is. If it is a problem that is recurring, and decisions need to be made every day, then obviously data latency is an issue. But if a decision only needs to be made once a year, and there is a few weeks' latency in the data, that may be acceptable.

In this regard, the industry a business is in is important, and this is part of the upcoming roadmap for further development of the AI Impact Model. A greater degree of latency may be a critical issue in, for example, the technology industry, where the rate of change is very high; on the other hand, latency may not be a big issue in the retail industry when, for example, dishwashers, dryers or other appliances are being sold and the rate of change is not that high. On balance, data latency is more important in some types of industries than others.

Companies are now looking at agentic AI approaches that tie the data layer very effectively to the AI model layer and then to the action layer. In these circumstances, you may actually be able to solve the data latency challenges by moving faster through the rest of the chain to make up for latency.

Another thing to consider is that risks differ substantially the use case and the nature of the data you're dealing with. The high levels of risk involved with old data in healthcare or financial data argue for very low latency. For example, in financial services, suspected fraud, and in healthcare, adverse events, should be surfaced and reported quickly. But there may not be as much risk associated with latency in industries in which rapid responses are not essential.

Risks

There are two aspects to risk in AI projects: ethical challenges and privacy and regulation.

Risks-Ethics

Ethical concerns, such as biases or unintended consequences in data usage, must be addressed to avoid moral, reputational and legal risks.

We have defined five risks of this type:

- Bias and discrimination
- IP and liability issues
- Economic equality
- Misrepresentation of data
- Environmental impact

In determining the risk of these issues, we could not create a five-point scale. Instead, our analysis is that risk is high if there are more than two ethical challenges out of these five. If there is no ethical challenge, the risk is low, and if there are one or two such challenges, the risk falls in between.

Risks-Privacy and Regulation

This subdimension measures risk in terms of privacy and/or regulatory concerns. It's a policy issue that relates to whether you can use certain types of data without getting permission or if you need to have the right data acquisition strategy in place. Using Facebook data or other social data feeds has implications under GDPR and other statutes. There may be regulatory considerations pertaining to cross-border legality of data exposure, etc.

The scoring depends on the nature of the data and the geographies in which it can be used. Europe has its own way of handling data with stricter norms than, for example, some developing markets, while the US falls somewhere in between.

These risks are table-stakes issues. Dealing with them is necessary but not sufficient for success. If you fail to deal with these issues, it can cause your project to fail, but if you do adhere to the applicable rules, it will not guarantee that your project succeeds. Privacy concerns can derail a project if not dealt with. Some level of privacy protection is typically already present in many data layers. The level of protection needed with regard to ethics and privacy depends on the sensitivity of the data, the industry you are in, and the kind of function being performed.

Not all data is equal in this regard; in some applications, you may not have data that relates to privacy or contains personally identifiable information. If you do work with personally identifiable information (PII) data, you should address it upfront in your data systems and handling; this is something that is often much more difficult to fix down the line. If you've exposed data of this type, there can be huge regulatory and compliance problems which can result in reputational issues for the organization.

If there is leakage of sensitive data, the payoff on the project could easily turn negative, given the deleterious impact on an organization of the legal or reputational costs associated with such issues. Given these potential impacts, planning to deal with any privacy concerns before project launch is highly recommended.

Scenario: Poor Data Quality and Excess Latency Dooms a Retailer's Quest to Improve Stocking Practices

Company Type: Retailer of consumer goods

Objective: Determine lost sales by region to improve stocking efficiency

Issues: Data quality and latency

This scenario focuses on a retailer aiming to use information related to lost sales by region to identify items that would require buffering or safety stock. The idea is to create a store of stock that the company could draw on during times of high demand. The company also wanted to identify items that were not selling quickly so they could be rationalized and taken out of circulation.

Similarly, the company wanted to check which markets it should target for increased stock. For example, demand patterns are likely to be significantly different between Harrisburg, Pennsylvania, and Manhattan, New York City. As a result, the two markets should be stocked differently. The company tried to understand from the data which products and which markets should be buffered by adding more stock in order to optimize its costs and at the same time capture anticipated demand and increase revenue.

When the company started looking at the data, it ran into problems. It could have used shipment data, but it realized that shipment data was internal data that didn't incorporate the impact of external factors. These could include supply chain issues, or how retailers were prioritizing their products compared to others or how the demographics looked in the region where the products were being sold. Thus, the company decided, "We need to bring all of that information together and then segment the markets and decide which segments we should be giving a higher priority, which products we should be giving a higher priority and stocking in excess, and which products we should stop stocking."

Both the availability of data and the quality of data in this scenario were poor. This retailer's data was managed by a third party. To get to the data took a few days. Thus, the company was unable to access the point of sales data as and when it was needed. When it did get the data, it was abstracted to a degree that prevented any useful conclusions from being drawn.

As a result, the entire project failed because of the company's assessment of data quality. They expected that the data would be integrated, and the integrity would be high. They also thought that it would be at the right level of granularity. That did not occur, leading to the cancellation of the project. The retailer was not able to find the right strategy and incurred sizable expenses due to both overstocking which ballooned inventory carrying costs and understocking resulting in significant delays.

The takeaway here is that when data availability is linked to other parties, it can pose problems in terms of both timeliness and quality. In this case, because it was relying on its partners for data, the retailer couldn't get the data as quickly as it wanted. Additionally, the quality of the data was not high enough to enable the retailer to benefit from the data when it was available.

Recognizing the Importance of Data in AI Projects

Our research has identified data as a vital pillar to AI project success, but in some cases, organizations fail to fully realize this. A primary reason for underestimating the importance of data is a tendency to believe that simply having access to great technology and models can solve business problems. The end result, or objective, of any AI project, of course, deserves significant focus in the planning process and so does using the right models and technology. However, this shouldn't detract from planning aimed at getting the functional elements right.

The factory analogy can be helpful here: if you're enamored by the car you're going to build, but you don't think about all the raw materials and manufactured components you're going to need to build the car, it doesn't matter how brilliant your vision of what the car will look like is. If we convert this analogy to AI, data is the raw material. Data, as is sometimes said these days, is the new oil. In fact, in our conversations with companies that have implemented AI, generally, sixty percent to ninety percent of the work needed to get the project off the ground involves getting the data in order. This takes time and effort and does not deliver visible ROI by itself. Thus, it often doesn't get prioritized enough.

While the models are a crucial element of a project, they generally don't require as much time and attention to build. The data, on the other hand, can be time-consuming to gather because it's often all over the place. It's in silos, it's not clean, it's not accurate, it's not unbiased, it's not current, etc. If data is not properly prepared for use with AI, it can cause significant problems, as seen in the preceding scenario. In many large enterprises, we've found that it can take a significant amount of time and effort to get their data in one place. They often have to pull together data across business units, across geographies, across functions. In some cases, pulling that data together can take years.

As a result of the difficulty that can be involved in gathering data, some organizations underestimate the cost, complexity and time that is needed to build a clean database. It's important to have a single source of data for these projects. Additionally, the data has to be cleaned, it has to be integrated, it has to be organized, and it has to be structured. It has to be made AI-ready.

People who are overly focused on the outcomes may not understand how much hard work goes into this. As a result, they tend to underinvest in the data infrastructure and staff training and the focus needed to maintain the integrity and the integration and the quality of the data. To avoid this, it's vital to realize that AI eats data for a living. If you don't give it the food–the raw material–it needs, how can you expect to get the desired outcome?

At the end of the day, AI is all about the data. This is as true for generative AI projects as it is for machine learning AI, as any outputs of these projects will only be as good as the inputs provided and those should include context from the organization's data. While generative AI can create outputs on the basis of public data it is exposed to, companies need output that is contextualized to their business. To accomplish that, it is important to augment the generative AI model with organizational data through techniques like Retrieve-Augment-Generate, Prompt Engineering, and Fine-Tuning.

With regard to agentic AI, embarking on agentic AI projects that use a weak data foundation is inherently very dangerous, as agents can take outputs straight to deployment and that could lead to negative outcomes if the agents are using unreliable data.

To ensure you have quality data, it is key to put in place a robust data governance system or framework. This includes a data audit that asks the question, "What data do we have and where is it?" It involves data stewardship, which centers on who is going to be responsible for managing and governing the data. It also involves policies relating to how data is collected and used. In addition, there are further elements that require attention such as privacy and data security. This comprehensive data governance framework comprises a combination of roles, responsibilities, data stewardship, and processes for managing data, along with the policies around the data.

Getting a data governance framework in place typically begins with an audit–which provides the opportunity to get a state of the data. It offers a chance to answer questions such as where is our data? Where is it sitting? What are the problems with generating and accessing it? Often, organizations create a data lake that is a single source of the truth, either virtual or physical. Virtual, in this context, means you virtually connect the different data sources into a virtual data lake. Physical means you actually build an enterprise-wide data warehouse where you merge all the data. Whichever approach an enterprise uses, the takeaway is that you have to spend the time and be mindful of the importance of data and make the necessary investments to ensure your AI projects receive quality data.

Another challenge centering on data is that there are three ways of classifying data. These used to be designated as velocity, variety, and volume. All of this is exploding. Now, just when you're able to handle all the structured data, you suddenly have to contend with unstructured data. Customers are talking about your organization on Reddit. They're talking about you on forums. You have social data, location data, telemetry, meeting notes, and various forms of internal and external communication. You have all these new forms of data. Now, your data platform needs to be able to ingest not only structured data, but all these disparate forms of unstructured data; additionally, all of them need to be put together before you can build a complete profile of your customers for AI.

For example, if an insurance company, let's say an auto insurance company, is looking at how to price risk, the question is, what do they know about their customers? They know where they live. They know what car they own. They know their previous claim history. They know how many miles they drive. They have their demographic information. As a leader of this company, you could put this in a simple spreadsheet for each customer. But now, you're able to put a device or a sensor or an app into their car and measure exactly how they're driving. You have data on road conditions, traffic conditions, driver behavior, weather, etc. And by the way, this data is coming in real time–it's streaming. Let's say a customer is driving 300 miles, and you're getting continuous data. This raises the key question: How in the world do you handle that much data?

You can use this data to make very precise AI models that say: Because this customer tends to accelerate like a jackrabbit, you're going to price your insurance higher. But making that prediction to an AI model is like opening a fire hydrant–you have to be able to drink from it. That is the other problem enterprises are facing; they have all these new sources that provide data at a much faster velocity, and now it's not enough to take that data and sit on it for six months and then price the insurance. Ideally, you turn on the car and your insurance starts getting priced. It's during the trip that things change; if you start accelerating like crazy, your insurance rate could go up during the trip. That requires you to analyze this data and make a prediction and then implement that prediction in real time or near real time. Processing data at that scale is a challenge that organizations must solve if they plan to successfully deploy AI for these types of applications.

Another issue organizations may encounter when gathering data for AI, as mentioned earlier, is the phenomenon of diminishing returns, which means that adding more data at some point may not be worth the effort it takes to gather it. A simple question you could ask yourself to avoid this situation is, what percentage of the data that we collect are we actually using? Also, ask end users of the data how difficult it is to get access to the data that they need. You might also ask your IT experts, with regard to an application, how easy or difficult it is to actually access the data for the application. Those types of questions can help you zero in on how big the problem is. Focus on the type of data in use or the end users of the data and their complaints.

Data availability is important and valuable. However, it is also important to get the right amount of data. As mentioned earlier, trying to collect every piece of data eventually leads to diminishing returns. If you wait for one hundred percent of the data to be available, the chances are that circumstances will have changed, and it won't provide the value that you're looking for.

Our guidance to enterprises is not to worry about having one hundred percent of the data available; it may be acceptable to start a project with sixty to seventy percent of the data to begin getting results. If you wait for the entire availability of data, it can become like a case of the chicken or the egg–waiting too long will generally not deliver results as quickly as needed. However, it is important that the data you are using is representative of the overall data set (e.g., if you are training models for your customers and data from certain regions has become available earlier, and you build your models for national deployment on data that is biased for certain regions and omitting others, that might cause issues as your models may be ineffective in the regions omitted).

"Having the right data at the right time is an important aspect of AI success. I think if you're waiting for the perfect data, you are never going to do any data analytics or AI, but at the same time, if you want to build trust within the business, you need to build your data quality and data governance frameworks so that you can make faster and better decisions. I think once you are clear on the business problem that you're trying to solve, you'll be able to figure out what actions you can take to get the right data and derive the knowledge and insights you need from it."

Deepak Jose – Global Data, Analytics and AI Executive. Vice President, Head of Data and Decision Intelligence, Niagara Bottling; formerly served in executive roles at Coca-Cola and Mars

As befits its importance to AI projects, we'll end this section by returning to the issue of data quality. This is determined by the variables discussed earlier in the chapter: how current is the data? Is the data properly integrated? There are numerous variables, and they talk to and influence each other to a degree. There are many horror stories about companies not being able to fulfill orders and the customer experience going south because of bad data.

Data quality is absolutely essential to the success of AI projects. Poor data quality can cost enterprises a significant amount of money. Research from Gartner found that poor data quality costs organizations $12.9 million each year across all industries.[6] Another study, from MIT Sloan Management Review, reports that most companies lose fifteen to twenty-five percent of revenue annually as a result of poor data quality.[7]

Scenario: Privacy Issues Torpedo NLP Data Mining Project

Industry: Multinational technology company

Objective: Convert customer interactions into business insights

Issue: Privacy breaches halt an AI program

This large technology firm experiences significant customer interactions, typically at its help desk, mainly consisting of how to use its software products and resolve different software problems. They executed a program using semantic technology and text mining technology via natural language processing to convert this into insights. They were able to use these insights in their product planning: what features to build, etc., depending on what queries they were receiving. That helped them drive faster deployment of new features and debugging. This made their product more desirable and gave them a substantial competitive edge in the market. This project guided their product planning and helped them become more competitive. They worked with an AI consulting firm to design and implement the project.

Initially, the program was very successful, and they started to get very good results, but over the course of time, someone within the organization from the chief data officer's department realized that data from all these customer queries and interactions was being sent to the vendor without masking that data appropriately. As a result, the personally identifiable information of the person who was making the query was also sent by mistake. This caused huge blowback, and the entire program had to be shut down. Having burnt its fingers, the organization didn't move forward with any aspect of the project, which had presented them with a significant opportunity to create business impact. They shut it down and for a couple of years didn't even explore it, because there was so much fear caused by the issues the project had experienced.

The takeaway here is that the issue that brought what could have been a highly productive program to a complete halt stemmed from a very simple step that was omitted in planning for privacy issues in the data used for the execution of the program. If this had been adequately planned for, and the masking had been done properly, they would not have encountered this issue at all.

The Danger of Waiting for "Perfect" Data

Data needs to be connected to the problem you are trying to solve, and it has to be secure, accurate, etc. But there is a somewhat counterintuitive finding that it is not necessary, and can sometimes hurt a project, to try to obtain the highest quality of data possible. This gets at the idea that the problem with data is that it is never perfect. You can always do better or, at least, think that you can do better. For example, let's say that you are getting quarterly data on automobile sales. You may think, "Well, I wish I had monthly sales." And if you're getting monthly sales, you may think, "Well, I would be able to react better if I had weekly sales." And then, "I would actually do even better if I had daily sales."

A scenario similar to this actually occurred to a global automotive company. In this case study, the company created a superior sales reporting system for dealerships that allowed them to give dealers daily sales data. And the result, rather than improving things, was chaos. Upping the frequency to daily reporting caused dealers to make knee-jerk reactions to short-term data, when that may have simply been a blip. The idea being that as you get to more frequent data, there is a point beyond which there are not only diminishing marginal returns, but it actually makes things worse.

What they learned at the end of that exercise was that the system only had the capacity to respond on perhaps a monthly basis. Giving the system daily data just caused things to spin out of control, as people on the ground could not calibrate their responses appropriately. There is a similar phenomenon when people start pushing to keep improving quality. How far do you improve quality if doing so ratchets up the cost to acquire data beyond what makes sense given the efficiencies you hope to realize from a project?

There may be diminishing marginal returns to how much you keep improving quality because doing so is not free. So, improving quality from 90 to 95, or 95 to 98 may be productive in terms of ROI. But when you get to going from 99.1 to 99.2. or whatever the threshold may be, then it starts to become so expensive that it's not worth doing. At that point, you have to be pragmatic and say, "Let's move on." The idea here is that while it's good to seek better data, at some point, it becomes too costly.

Additionally, your ability to process that level of information is also constrained. As a result, there is a stopping point beyond which it's counterproductive to keep looking for the perfect data because not only is there no such thing but also it simply consumes more resources than it's worth. This idea can be counterintuitive because if you ask somebody, "Is data quality important for AI?" the answer will almost certainly be, "Yes." But the issue is not whether data is important in general. That is well established. The issue is, "What level of data quality should we have?" If the answer is the absolute best, most perfect data, we risk making the search for perfection the enemy of progress. There are, of course, exceptions to this general rule as there might be certain use cases that require absolutely perfect data.

Scenario: How Poor Data Quality Almost Derailed a Predictive Analytics Project in the Banking Sector

Company Type: Large banking institution

Objective: Use predictive analytics to inform targeted marketing campaigns

Issues: Difficulty with data quality caused by scaling from small test data sets to production level quantities

This scenario demonstrates the crucial importance of focusing on data quality when launching an AI project, especially during the scaling process. The customer, a banking firm, wanted to use its data on customer preferences, assets, and past behavior to identify those customers most likely to purchase specific products. For instance, customers who had previously purchased certificates of deposit (CD), had expressed interest in learning more about the bank's CD offerings, or had high cash balances were predicted to be most likely to buy CDs.

Another example along these lines is related to retirement products such as individual retirement accounts (IRAs) sponsored by the bank. By tapping into its database to track customers who were of a certain age, had a certain amount of assets, or had expressed interest in learning more about retirement products, the bank hoped to more precisely target its marketing efforts for these products.

The predictive analytics program used by the bank estimated that using these insights about customer buying patterns could generate significant ROI by enabling the bank to aim its marketing efforts at only those customers most likely to be interested in the products being offered. Relatively small data sets were used to test the program in an isolated environment designed to replicate the customer's operational IT system. In these tests, the bank was able to generate accurate data based on the specific product being marketed. After this, the go-ahead was given.

Problems occurred immediately upon launch. The issue was not with the assumptions underlying the use of predictive analytics, but with the quality of the data generated by the software utilized to implement the program. While it had performed well in small tests in an isolated, sandbox environment, when launched at scale in the bank's IT infrastructure, the process of drawing information from the disparate software solutions where the data was housed resulted in the production of inconsistent data. Customers who had expressed an interest in one type of product were incorrectly classified as having expressed an interest in a different product; additionally, in some cases, the system failed to properly categorize customers by assets.

This type of scaling problem is not uncommon. Programs that produce accurate results in an isolated test environment may malfunction when placed into the customer's operating IT ecosystem. Sometimes, exponentially larger volumes can lead to errors in processing for technical reasons.

To salvage their investment, the bank formed a team of internal and external experts to diagnose where the system had gone wrong in categorizing the data used to produce lists of prospective purchasers of the relevant products. Working around the clock, this team was able to identify the root cause of the issue and patch the software in question so that, at scale, it no longer misidentified which customers fell into the categories identified by the predictive analytics program. With the quality of the data ingested by the program now at a high level, the bank was able to proceed with its targeted marketing campaigns. These campaigns, as predicted, realized a higher ROI than past efforts which had not used this AI model.

The lessons learned in this scenario center on the importance of ensuring that the data used to power an AI program is of consistently high quality. The predictive analysis algorithm at the heart of the bank's project was capable of generating actionable insights, but the difficulty of pulling data from multiple systems caused errors to creep into the process of turning those insights into accurate customer lists for marketing purposes. This highlights the importance of a robust testing process that monitors output through all phases of a project to verify satisfactory data quality each step of the way.

Building Bridges Between Data Types

The key to AI project success is to generate data that is of the highest quality with reasonable effort given the scope and constraints of the project and make the most of that data. One way to do that is to build bridges between data from different sources. This is another very interesting insight we've discovered. For example, let's say that you're trying to model the productivity of an AI program based on three different types of inputs. Using the example of a wireless company one of the authors worked with, data was available from a variety of sources at the business. There was customer satisfaction data, which consisted of what customers had to say about the product. You also had data from marketing relating to things such as what price promotions and plans a particular consumer was responsive to. And you had data from network operations saying how good the coverage was in a particular area. As it turns out, there is significant value in connecting these different types of data.

A relevant quote here is that *innovation happens at the grout between the tiles*. The idea being that the more different types of relevant data you get, the better your insights will be. In this case, it's like triangulating on the customer, and if you're shining the light on four or five different angles, you get a much richer picture of what is happening. AI models become richer when you're able to say, this customer satisfaction is lower because they live in a zip code where our coverage is not great; and by the way, as a consequence of that, if we give this customer more data next month as a promotion, that doesn't help them because our data quality and our network coverage is poor in their area. So, maybe we want to give them some type of discount or a better device or a hotspot that they could put in the home or a cellular antenna.

Now, if we imagine that we chopped off one leg of this triad of data, what would happen? The model's three legs are customer data, pricing and promotion data, and network data.

Let's say that for whatever reason you don't have access to the network data. You could now invest a lot more effort in getting more precise about customer satisfaction data. You could now use Yelp and similar sources. You could try to do more surveys. In other words, you're investing more in the increased quality or improved quality of the customer satisfaction data, but you're only operating with two legs. Our contention that is there is much more value in activating the third leg, even if all three legs are not perfect, than to try to perfect one leg and stand on it. That is the value of unifying different types of data. You get richer insights when you're able to make connections across different aspects of the business, and different aspects of customer behavior.

CHAPTER FIVE

Technology: Enabling Scalable Solutions

"In general, when it comes to technology, I think organizations should have an adaptability mindset as opposed to what is called building to last. This gets at the concept of building to last versus building to adapt. In the past, building to last was, for example, saying, 'Let me buy this software, then I won't have to change anything for twenty years.' With today's rapid rate of change, that is out. Now, building to adapt is very important; as a result, understanding the role of technology and especially AI in your business continues to be vital."

Deepak Jose – Global Data, Analytics and AI Executive. Vice President, Head of Data and Decision Intelligence, Niagara Bottling; formerly served in executive roles at Coca-Cola and Mars

Technology is the enabler that transforms AI programs into actions or actionable insights. At the same time, selecting the right technology stack is often a daunting challenge. This chapter examines the third Value Lever–Technology–highlighting the role of orchestration tools, insights platforms, and integration. It explores the pros and cons of best-of-suite versus best-of-breed approaches, helping readers make strategic decisions about their technology stacks.

A case study covering a CPG company's Strategic Revenue Management Toolkit demonstrates how the right technological infrastructure supports scalability and agility. Readers will gain practical advice on evaluating their current technology environment, aligning tools with business goals, and investing strategically to enable seamless integration and scalability across the enterprise.

The chapter focuses on the two major systems involved in planning and implementing AI projects covered in the following sections:

- **Orchestration Technology**: Tools and platforms that enable seamless integration and coordination of AI models to drive engagement and action.

- **Insights Platform**: Availability of a centralized system for data visualization, analysis, and insights generation for decision-making.

Orchestration Technology

Orchestration focuses on the other tools required to drive action. The first question to ask is whether those tools are available and in place at an enterprise or not? For example, do you have access to Python libraries? Do you have access to cloud infrastructure? Do you have data processing engines and data modeling engines or MLOs (machine learning operations) in place? Do you have the core platforms that you need to drive the actions that are prescribed by your AI models?

In the coming months and years, orchestration technology will become much easier to obtain and deploy. Things are progressing so rapidly in the AI world these days, with foundational AI technologies and agentic AI, that access to these types of applications is increasingly common. This wasn't true just a couple of years before the writing of this book.

Startup companies are now building orchestration tools that are designed to hyper-scale. While that may be the future of these tools, dominating the market currently are orchestration suites such as Azure Open AI Studio from Microsoft and Salesforce with Agent Builder. These can basically be used to build all your agents. However, in doing so you are locking yourself into a vendor-specific orchestration framework.

It's similar to how Apple builds its ecosystem with iMessage, iPhone, and iCloud, making it very hard to leave once you've spent some time there. If you start to use Google's orchestration framework, Google Cloud platform, then you become a Google customer, or if you become an Amazon customer, they have Bedrock. Thus, even at the orchestration level, you're locked into a specific vendor. The advantage is that implementation becomes very seamless. The challenge is that they don't make it easy for you to orchestrate across third parties, and they set it up so that if you work within their stack, things work better. However, there are also substantial benefits to using one of the large-scale platforms in terms of ease of use, linkages and scalability, so it is really a choice that the business has to make to tradeoff between the advantages and disadvantages of either approach.

You need an orchestration framework because without one you can't get these agents to talk to each other and these AI tools to talk to each other. However, the question is, again: can you do this in a way that is platform agnostic or vendor agnostic as opposed to getting locked in? In terms of a recommended strategy, it will always depend on how important integration is to your organization, and how much investment you already have in your existing systems. If you've already decided, for instance, that you will be with Salesforce for the next ten years, then perhaps it's best to go with their AI suite. On the other hand, if you are building from scratch, and you don't have a huge investment in existing infrastructure, then you may want to put together a best-of-breed framework and put your own orchestration on top. The decision really depends on how much legacy infrastructure you have and what your priorities are going forward.

Is your priority that you want fewer potential headaches from system integration or that you really want to be on the cutting edge of the technology? For example, if you run a marketing services company or an ad agency, perhaps you need state-of-the-art tools for content creation because that's your core business. In that context, you might choose to go with the best-of-breed approach. But if you run a retailer that just wants content for your catalog, and you already have your database sitting inside a CRM from Salesforce or a similar vendor, you might choose a best-of-suite approach and go ahead and use their tools.

Insights Platform

There are three systems that create data to drive actions. The system of records is where transactions are recorded; the order comes in, you take the order, which is then recorded by the system. The system of records will show levels of inventory outstanding orders and production schedules. Typically, an organization's managers will look at these various data points and use them to order raw materials, schedule production and set prices; these actions will typically take place in systems of engagement or action. In recent times, organizations have built up "systems of insights" which enable managers to choose "actions" more intelligently by using descriptive, diagnostic, predictive and prescriptive insights regarding the data available in the system of records and system of action or engagement.

At different frequency levels, you start making sense out of the data. That means you start creating insights. In the past, an organization's system of records used to provide all of its system of insights as well. It used to say something like: "This is what has happened in geography X for product Y." It would provide, after analysis, all the descriptive analytics necessary to answer questions such as: what happened? When? Where? How much?

Over time, evolution in that process led to analytics moving from descriptive to becoming more diagnostic. If things have gone wrong, why have they gone wrong? This causes analysis to become predictive because you now know what happened, what went wrong, why it went wrong, etc. Therefore, you have a better grasp on what's likely to happen. This is forecasting, which utilizes predictive analytics. The next step is prescriptive analytics, which focuses on answering the question: what should you do? What steps should you take? All of these functions are now part of a system of insights.

Typically, the maturity of a system of insights is determined based on the level of insights and analytics you are performing. Are you only focused on descriptive analytics such as what happened, when, where, and how much? This is at the lower end of system potential. Higher-end systems add in diagnostic, predictive, and prescriptive capabilities as to what's likely to happen. AI can now be applied across this entire value chain and, if used effectively, the process of getting descriptive, diagnostic, predictive and prescriptive insights can take place on steroids.

A system of engagement is typically where the final actions are executed. An example would be a salesperson in the retail world at a beverages company who goes to a retailer and sees that the company's products are running low on the shelf. So, the salesperson then uses a device to say that packs of a bottled beverage product, for example, are running low, and asks for replenishment from the warehouse. That device and its backend system are part of the system of engagement.

That system may also consume insights that have been calculated from somewhere else; for example, saying that certain products are likely to run low. That consumption of insights and then being able to take actions based on them occurs in the system of engagement.

In some cases, all three can be the same. But as time has progressed, we are finding that there is a partition occurring between systems of records, systems of insight, and systems of engagement. Sometimes a system of records is actually also a system of engagement because you can take an order and take actions as well; for example, the SAP system, which is an ERP and a system of engagement.

The system tracks when transactions occur and is linked to devices that will allow you to take data from systems of records–such as the example above of the system saying packs of soda will be low–and then letting a salesperson use the device to engage with various parties. Systems of engagement and action need to be built for scale in ingesting input and allowing for downstream action. This means what's happening on the system of insights is available in the system of engagement and is also connected to the system of records.

The insights platform is as much about having the right technology investment as it is about the person at the point of the decision having timely access to it to support their decision-making process. This may be someone on the shop floor or a salesperson in the field for an operational decision or a senior executive for a strategic decision. A high-quality insights platform is also about having the right linkages to the sources of relevant data which are published with supporting analysis in a dynamic and timely manner to the stakeholders. Now, with generative AI technology, it is not enough to just have BI platforms. Natural modes of consumption where users can interact with their data via text or voice and get the insights they need on the fly are also needed.

Case Study: Choosing the Right Technology to Support Scalability and Agility

Industry: Consumer packaged goods (CPG)

Objective: Improve decision-making with an SRM system

Issues: Traditional information gathering process doesn't support timely decision-making

This case study reveals how a CPG company's strategic review management tool kit demonstrates how the right technological infrastructure supports scalability and agility. A company in the CPG sector was seeking to improve its ability to quickly react to changes in its markets by implementing a strategic revenue management system utilizing a sophisticated tech orchestration system. Their system integrated their system of records with their system of insights and orchestration system to enable rapid decision-making when applicable.

This case study focuses on the intersection of SRM and technology orchestration. The company's goal was to set up the tech infrastructure required to support the scalability and agility needed to make rapid decisions when called for.

The company used a strategic revenue management tool kit, which consisted of code that enabled them to perform pricing simulation and pricing diagnostics. This tool kit provided them with a competitive advantage. This enabled the company to react to changes in a timely manner. The key to that was to set up a platform that facilitated getting this information into the hands of the decision-maker. That enabled them to simulate how much revenue they would realize if they took certain steps.

Revenue management typically has four or five levels, or pillars: pricing, promotion, trade spend, pack price, pack architecture or packaging. These are the key levers. Thus, if you can simulate very quickly what would happen if you were to change your price and give a discount of ten percent–what the remaining volume would be and what the impact on the revenue would be or what the impact on margin would be–you can gain a competitive advantage. In this case, the company used this technology to very quickly determine the right price points to get the maximum revenue uplift.

They used the AI program to avoid the typical, much slower approach of sending in a series of questions and having somebody else do the analysis. Then, after two or three days or more, it would be sent back. In this scenario, by the time you make the decision, your competitive advantage is reduced. The takeaway here is that putting the power to simulate in the hands of decision-makers enables nimbleness and speedy decision-making. The company accurately defined the crux of the problem as timely decision-making and enabled this result by achieving alignment among leadership that this project was worth supporting. Additionally, the proposed project was relevant to the business users, and trust in the system was high because it provided a comprehensive view of the data.

The project also scored well in the usability category as people were able to use it effectively after receiving a short training course. All of that, combined with effective integration between disparate technology systems, enabled the project to deliver to decision-makers–in a timely manner–data of the necessary granularity to make the project a success.

AI Strategy and Systems of Insights

As described, a system of insights can have descriptive, diagnostic, and predictive/prescriptive types of analytics, and AI typically comes on the latter end of the spectrum, which is prediction. Prediction could be for any aspect of the business, whether demand forecasting like machinery failure. Prediction occurs on the basis of the data that is available. For demand forecasting, you will need historical sales data, pricing data, promotion data, not only yours but your competitors and so on and so forth. With agentic AI, you can also move from just predictive and prescriptive to taking cognitive actions directly from the prescriptions of the AI.

The key question for a system of insights is: do you have your insights platform that provides the insights linked with the system that creates actions and integrated with your engagement system so that you're able to engage with the output of AI?

When enterprises embark on an AI strategy, an important insight we're finding in this model is that it is important to make sure that your systems of intelligence–your AI models–connect back to your systems of records, which are your enterprise systems, your ERP, CRM, and so on. The reason is that they have to pull the data from those systems. Those systems are recording what's happening in your enterprise and then the AI makes sense of it and adds a decisioning layer. As covered earlier in the chapter, all of the systems of insights are now evolving from the old world of BI–business intelligence–to these new more cognitive systems.

If you're not able to make those connections, you cannot get AI models to scale. You can do pilots, you can do little experiments by pulling out some data from a system of records in a sandbox type of setting and then playing around with it, but that's not going to scale. The reason is that the system of records is telling you what happened, while the system of intelligence, which is where all the AI sits, is telling you why it happened, what might happen next, and what it means. The deeper, the more robust your connections are between these two sets of systems, the better your insights are likely to be, and the more productive your AI is going to be. The challenge is that these two systems sometimes don't talk to each other–they're not friends with each other, you might say. The systems of records are often legacy systems that have been around a long time, but now you need to get that data into the system of insights, or intelligence, to be able to analyze it.

So, what we're finding in the model is that companies that do a better job of connecting those different types of systems will get superior results. This means that even if you have a perfect AI model and a perfect AI system, if you're not hooked back into your enterprise data, your system of records, then you're not getting full visibility. You're not getting a 360-degree view of what's happening in the company. On the other hand, if you have a good 360-degree view of what's happening in the company, but you can't perform the analytics around it, then you're lacking in insight. So, these two things need to be connected, ideally, via deeper integration.

As a relevant analogy, consider a famous saying which has been applied to the city of Istanbul, which sits surrounded by water, astride the Dardanelles Strait: *water, water everywhere, nor any drop to drink*. The water surrounding the city is, of course, salt water, which does no good for residents of the town looking to satisfy their thirst. It's the same with data. Unless it can be made available in a form that your system of intelligence can analyze, it doesn't matter how much data you have or how high in quality that data is–it's useless to you from the perspective of AI analysis and implementation.

The question is, what is the best way to make those connections from the AI layer to your application layer to the data layer? One approach is to work with the vendors who are actually providing you with the system of records. As mentioned in the previous section, Salesforce has its own AI suite, and it has Einstein; Microsoft Dynamics also has its own AI, as do Oracle and SAP.

The argument you might hear from a Salesforce a Microsoft or an Oracle salesperson is that our AI suite works much better with our system of records. That is the best-of-suite strategy. It is based on the concept that a known system is better than an unknown one. Someone might say, "I have all my investment in Salesforce, all my CRM data is there. So, why don't I just go with them for the AI tools?" In that case, you don't have to worry about integration across these various layers.

As we have discussed, this approach–best-of-suite–provides the benefit of seamless integration; one vendor is responsible for all of it, you already have an investment in the infrastructure, and you just add the AI capabilities to it. But the challenge that this strategy brings with it is that, as the saying goes, a jack of all trades is rarely the master of all trades.

There are companies that build AI tools for a living. They build cutting-edge systems of intelligence and, arguably, since they're specialized, the state-of-the-art technology that you'll get from them is better than what you would get from the best-of-suite AI offerings. The best-of-breed approach is to say, "I'm going to look at each functionality in isolation." For example, if you want to write marketing copy and use blog posts in your marketing, you could use Salesforce Einstein as a content management system. Alternatively, you could go with a best-of-breed approach and use certain standalone tools, which are specialized and just do marketing content and marketing content management. The advantage of the best-of-breed strategy is that you're typically getting the state-of-the-art applications in each category.

The vendors who provide these specialized solutions are extremely focused on their niche because that's all they do.

The downside of the best-of-breed approach is twofold. One is that you now have to do the integration yourself and, if you're not careful, you end up with a menagerie of tools that you now have to take the time and trouble to manage. Additionally, sometimes it can be more expensive because now you're paying multiple vendors. It's like adding more subscriptions to your OTT.

The second challenge is that, by definition, some of these newer companies are startups, and are therefore not very stable. Thus, you face what is called stranding risk, which is the risk that they go out of business, or they get acquired, and suddenly you're left in the lurch. You may have all your data sitting inside one of these solutions, but if the company providing the solutions goes bankrupt, now what? Salesforce, Microsoft, and Oracle are large, established enterprises and highly unlikely to go away. If that happens, if you bet on going with something that was better, but then they left you hanging, you now have to deal with it. That is the tension between these two approaches. Ultimately, you're solving for two things:

1) Better integration
2) Better functionality

Best-of-breed gives you better functionality, while best-of-suite gives you better integration, and how you navigate between those two objectives will determine what type of AI system you ultimately select.

We also talked about best-of-breed and best-of-suite in the deployability section of Chapter Three. These concepts are closely linked to decisions surrounding orchestration technology and insights platforms. Essentially, deployability talks about connecting across systems, while the orchestration technology is focused on the quality of the end system in itself. You have to have the right technology and that depends to a great degree on the use case–on the problem you're solving.

In some cases, you're providing insights to a person to enable them to make a decision. The key for every insights platform is to get the right information to the right person at the right time.

In other cases, you're deploying a model directly into a system that is taking an action, so the orchestration technology is crucial. These two things differ in their need and value depending on the users involved.

Scenario: Technology Orchestration and System Insight

Industry: Information Technology

Objective: Improve customer service

Issue: Bring together data from disparate sources to quickly answer customer queries

This scenario revolves around the integration of deployment at scale. In this case, there are customer service representatives who engage with customers and solve their issues or problems. A customer service representative from a PC OEM technology firm is responsible for figuring out how to ensure that its customer experience and customer service are top-notch. The representatives have targets to measure, such as turnaround time or resolution time, if a complaint comes in, based on how quickly the order can be taken, responded to, and fulfilled by setting a certain schedule. The company typically receives a call from a customer saying something like, "My PC is misbehaving and not functioning properly; I want you to solve this because I'm under your one-year warranty."

The first thing that has to happen is that the system the customer service representative uses, which is most likely the system of engagement, will allow them to take that complaint and do some troubleshooting. This allows the rep to identify and hone in on what the problem really is: Is it because of some software glitch? Is it because of a hardware issue? Is it because of the internet? Or perhaps the person has forgotten to plug in the device, and it ran out of power? They have to figure all of that out.

This illustrative scenario demonstrates how technology deployed at scale that integrates systems of records with systems of insight and engagement transforms customer service in the PC OEM industry.

Imagine a customer calling under warranty with a malfunctioning PC. Traditionally, service reps spent days logging complaints, manually troubleshooting, and reconciling data across disparate systems. Today, with integrated **systems of engagement, records, insights, and automation**, the process is streamlined and far faster.

- The **system of engagement** captures the complaint and enables initial troubleshooting (e.g., software glitch, hardware issue, or simply a power problem).
- A **knowledge management tool** helps the service rep (or chatbot) quickly narrow down the cause.
- The **system of insights** recommends solutions, drawing on past cases to identify the most effective fix.
- **Automation agents** handle repetitive tasks–logging complaints, scheduling fixes, or even resolving simple issues.

This integration reduces resolution times from days to hours or even minutes, significantly improving the customer experience. Chatbots powered by generative and agentic AI can now assist or, in some cases, even take over roles once done only by humans, providing consistent, fast, and adaptive support.

Unlike traditional **RPA**, which only executes repetitive tasks, **agentic AI is adaptive and self-learning**. It references past resolutions, leverages AI insights to recommend the best course of action, and learns from mistakes to improve efficiency over time. For example, if one approach took six hours but another just two, the system learns to prioritize the faster path.

The key enabler is **integration and deployment at scale**–ensuring engagement, insights, records, and automation systems work together. This allows service teams (and increasingly autonomous agents) to resolve problems quickly, raise customer satisfaction, and continually improve outcomes.

CHAPTER SIX

Talent: Building the Right Team

"To improve the chances of success for an AI project, go to people who have a curious bent of mind. People are very different. Some people are very good operationally. Some people like to solve problems. Some people just say, 'Well, that's not my thing.' You want to find a group of people who are fit for certain types of things. So, go find the most curious, most energetic people who feel excited about a new feature or about testing a new product."

Ajit Sivadasan – Ex VP and GM,
Global eCommerce, Lenovo

The return on AI projects depends critically upon the people who design, implement, and execute them. The fourth Value Lever–Talent–focuses on the skills of the people behind AI initiatives. This chapter explores the roles of technical experts and citizen AI scientists, emphasizing the importance of combining deep technical expertise with domain knowledge and business acumen. Readers will learn how to build and sustain high-performing teams through strategies for upskilling, collaboration, and external partnerships. A case study featured in the chapter illustrates how the right mix of talent and leadership drove the successful adoption of AI initiatives.

In this book, we divide the people working on AI projects, in one capacity or another, into two broad categories:

- **Citizen AI Scientists**: Nontechnical users who can leverage low-code and no-code AI and analytics tools to solve business problems without deep technical expertise. Citizen AI scientists (CASs) can be the bridge between business users and practitioners.

- **Technical Experts**: Skilled professionals with advanced knowledge of AI, machine learning, and data science capable of building, deploying, and maintaining sophisticated AI systems.

We discuss these categories of AI talent in greater detail in the following sections.

Citizen AI Scientists

The need for citizen AI scientists differs by organization. To some extent, it depends on another attribute, which falls under the Execution Value Lever, which is AI adoption culture. If your organization already has a substantial AI adoption culture and people have been using these tools themselves and they are familiar with the process, that makes them more effective when introducing these projects. Perhaps some of your businesspeople have become translators of these skills, and maybe they have become citizen AI scientists themselves.

There are other cases where such people are less common in organizations, and they could use more CASs to optimize their results from an AI project. There's a relationship between these two factors. It really depends on the state of the organization and what is already embedded in the organization's technology culture.

To support these projects when you're early in your AI journey, people who perform the functions of citizen AI scientists are definitely needed. The people who play this part may not be defined as citizen AI scientists–it's not a formal job title–but they need to have the ability to serve in this role. Whatever role a person is hired for within the business–it could be a customer support position, marketing manager, etc.–there are some people who, whatever they are called, can basically be identified as citizen AI scientists who can run these systems or apps. Having a few people like that on every business team enables the organization to drive AI adoption forward more meaningfully and seamlessly. They set an example for their business colleagues, support others when they get stuck, and serve as the "bridge."

With regard to the question of whether an enterprise is better served by hiring technical experts to implement an AI project or completely outsourcing this function, we generally recommend that it is valuable for enterprises to have a layer of technology expertise within the organization itself. This doesn't need to be a large contingent of employees. For most firms, having a small layer of technical expertise within the organization is very important because it gives the client organization an appreciation of what is doable. What is possible? What is working? What is not working?

They don't need to have the scale to do everything themselves, but just to understand the technology involved and correlate the needs of the business with the right technology approach to solve the technology problems it faces. They need to be able to dialogue effectively with external consul-tants and AI firms to ensure that the implementation within the organization is smooth.

In terms of working with more sophisticated implementations or when things are being built to scale, this often should be done through vendors, in our experience. We have found that some enterprises can do a larger part of this on their own, and some less. An organization that does not have a very well-defined technology roadmap and agenda may need more assistance from vendors than one that is further along in the journey. We recommend that, to the extent special expertise is needed for these projects, working with external partners is a good solution.

Additionally, when vendors work with various clients, these vendors bring significant learning, context and knowledge that can be used by multiple clients. This stems from their access to techniques and approaches that an organization would not otherwise be able to access. It comes down to core aptitude versus noncore. For example, if we consider Amazon, when it comes to defining cross-sell opportunities via an algorithm, they may consider this a core function and do it in-house and that goes to Amazon's business model. But for another company, let's say an apparel company selling various types of clothing, their core is their brand, their product; the upsell cross algorithm may not be in their wheelhouse because that's not their primary business. Thus, they are likely to look to outsource to gain that capability.

Citizen AI scientists are typically folks on the business side of the equation. They are the people who run the business, plan for the business, and understand its functions. Then, there are people who are practitioners, who are doers, who enable the use of technology and stand-up solutions for the businesspeople to make decisions. Those are the technical experts. We are placing citizen AI scientists on the business side of the equation.

Does the business team that is going to be the benefactors of the program have the required number of citizen AI scientists? In past years, you would have had two camps at a company: the business folks and the IT folks. Businesspeople did not care or appreciate the intricacies and difficulties faced by the IT folks, and IT people didn't necessarily care about the business side of things aside from technology. However, in the last several years this has changed as technology has advanced.

Now, the businesspeople may have enough training in AI tools to operate them and come to certain conclusions themselves rather than relying on IT folks.

The technology helps because it has become much easier for a person to get insights without being a programmer. For example, in the past, when business folks would ask, "Why is this machine likely to fail?" somebody would extract the data using a certain algorithm, perhaps called 'remaining useful life.' They would go to their side of the equation, solve for that and bring it back to the businesspeople.

Now the technology is there for the businessperson to say, "Hey, why can't I take this data, which I have access to already anyway, and use this tool? It will create the remaining useful life or automatically select what is the best tool to use, and then, spit out an output that I can use, and then I'll try to make sense out of it."

That's where this is trending, and citizen AI scientists are people from the business side who are open to learning and experimenting with the technology. An example from real life is Indra Nooyi, who was the CEO of PepsiCo Foods. Some years ago, she actually spent the time needed to learn programming and Python, as if to demonstrate to her colleagues that, *if I'm the CEO of the firm, and I'm doing it, you also better do it because that's where the world is going*. She probably didn't do it to actually write code for the company; she more likely did it to get a better appreciation of how it was done so that she could direct her teams more effectively.

Given the rate at which the technology has evolved, people with a non-technical background who are smart, curious, and bold can learn quickly from publicly available resources and get themselves to a space where they can function at a citizen AI scientist level. This is the most needed role today. There is, of course, still a need for a smaller number of "experts" who can set up data systems and build and manage complex algorithms, but by becoming a citizen AI scientist, one can create concrete value adds for the organization and for one's own career.

It's getting to the point where there is less need for people who code because coding gets created or generated on its own. There are a lot more citizen AI scientists these days. So, from no CASs available to having many CASs that can handle complex problems and envision solutions, that's the spectrum from very low to very high, and everything in between.

Technical Experts

If a firm doesn't have the technical wherewithal within the team to support an AI project, it will score low on this sub-lever. It ranges from no expert is available to an expert is available through hiring external consultants only, to some internal expertise available in pockets with support from external expertise. For a very high ranking of 5, internal experts are available and also well supported by partners. Smaller companies can also adopt alternative approaches–like retaining individuals on a fractional basis to design and manage their AI initiatives.

These are the people who can advise on the data strategy, including data governance, set up big data systems, build and maintain algorithms and understand how things connect at the technology level. These resources are also important to ensure that risks around privacy, security and ethics are appropriately managed and mitigated. These resources are typically expensive and difficult to hire and retain, especially for a non-tech organization. Fortunately, having a small team of these resources is sufficient, as that team can typically be effectively supported with external partners.

Building the Right Technology Team

With regard to building the right tech team, whether for integration or just deploying an AI project, this question arises: what should the nature of the teams that work on these projects be? Before we talk about upskilling, it's important to focus on skills in general. You can't talk about upskilling without knowing what skills are needed. To run or implement and deploy an AI initiative, you need a cross-functional team.

The cross-functional team has to have a number of specific skill sets. One, it needs to have a domain expert or somebody who understands the business and the workflows. They are typically from the line of business being targeted by the AI initiative. For example, if you're building an AI application for marketing, that would generally be a marketing executive. The second kind of person you need is a data scientist/AI expert. They're the ones who understand models, algorithms and platforms, so they'll actually be building the AI tools. IT systems expertise is the third kind of expertise your organization needs because, as we've discussed in earlier chapters, you're going to connect the system to your core enterprise systems–IT understands these systems. And the AI team understands the AI model. In order for that connection to work, you need representation from your enterprise technology people–your IT team.

"It's all about how we have leaders who are curious? How do we set the stage to ensure that there's a culture where you actually make learning a part of a ritual where you are constantly building this learn-it-all culture; you're building what I call the AIQ, which is the artificial intelligence quotient. That's something I believe is vitally important. And giving leaders and teams permission to experiment safely is also important because I think sometimes you have to create psychological safety, as we call it in the human world; you have to bring that similar kind of safety culture even in the AI world."

Ram Iyer – General Manager, Windows Devices & Sales, Marketing, Microsoft

We could add a fourth type of expertise, and that is project management. Somebody's going to manage the project and make sure that all the ducks are in a row. In a sense, it's almost like you're building a product–each of these initiatives is like building a product. Thus, it can be valuable to have a product manager who orchestrates, who connects, who makes sure that you have a roadmap and everybody's on the same page and you're executing, etc.

That's one piece of advice: to make sure that you have what might be called an AI squad or squads with eight to ten people–teams that possess these kinds of expertise. One further type of expertise that you might have on these teams consists of external consultants or vendors who bring in added experience and knowledge. For instance, if I'm working with the Salesforce platform, then I might also have a solution architect from Salesforce helping out.

Case Study: Building the Right AI Team

Company: Global technology company

Objective: Support talent to optimize AI adoption

Issue: Staff unfamiliarity with AI functionality hinders adoption

After some early challenges in launching AI projects, this company found the right mix of talent and leadership to drive the success of these initiatives. These projects typically feature many moving parts, among them the people aspect of the equation. The company found that even if a project was well designed from a technical standpoint, unless the people using the technology were prepared for the changes brought about by these projects, it could result in a project failing to achieve its intended results.

This issue applies to the initial stages of the automated insights project mentioned earlier in the book as well as other projects. The focus was on the technology side of the equation, and no one was keeping tabs on how people reacted and adapted to the changes to their work routine and responsibilities caused by the projects. What the company's leadership realized was that they needed to have change management for these initiatives, focused on ensuring that people understood the changes being brought about and how to work with those changes.

Once this change was made at a companywide level, the automated insights program and other projects became much more successful. ROI improved as the company was able to combine the technology advantages of AI with an improved AI adoption culture to optimize the business impact of launching AI projects. In investigating what was hindering the effective adoption of AI programs, the company's leadership discovered that roles for the projects had been assigned without making sure that the people involved in implementing and using the programs had been trained in their use. As a result, the changes being driven by an initiative were not fully understood by the people adopting it.

To remedy this, the company adopted a policy of training people in how these initiatives would change how they worked and make things simpler and faster for them. This led to the improved performance of multiple initiatives as the new policies took hold.

The insight that led to the improvement was that people didn't have the right training to fully implement these initiatives. They didn't have the correct training in this specific knowledge that they needed. Once that training was imparted, improvement was immediate.

They said, "Here's what this initiative will need in terms of training. Let's make sure the team is trained instead of just throwing them on it without the proper training." In training these people, they created a number of citizen AI scientists. That means they were now capable of doing certain things by themselves and not relying on technical expertise from others. Therefore, when you go on to the next project, because they have become citizen AI scientists and understand how to use tools and technologies, they are able to adapt more rapidly to new changes in technologies and other projects.

Upskilling for AI

There are generally going to be two types of employees in your organization, at a high level, with respect to AI. We can call them the creators of AI initiatives and then the consumers. Obviously, the number of consumers is much larger. They comprise the entire organization, in some sense. Then, there are the creators, who are going to be a smaller set. From an upskilling perspective, let's talk about these two groups separately. When planning upskilling related to the creation of AI initiatives, it's important to ensure that you have these elements of expertise: you have the right AI technical experts, you have the right domain experts, and those people understand the workflows involved and so on.

There are AI engineers, people who really know how to work with the actual models and the infrastructure, and in recent times, we have developed a new type of skill called prompt engineers, who are adept at using generative AI models. But more importantly, from a change management standpoint, it's vital to talk about the other set of people, the ones who are going to be the day-to-day users of your AI system. These are the consumers of its output.

Imagine that a creator creates an application that builds a personalized learning pathway that matches the employee's desired skills with their learning. It steers them to the relevant courses or learning material and builds the learning pathway. That's our hypothetical AI project, creating a sort of IDP (individual development profile). The AI matches the courses they could take with what they need to do on the job, so we need to find the right pathway for them. But then the flip side is that you now need to have these employees actually use and engage with this platform or app. And you need their managers to convince them that their success and advancement will be enhanced by taking these courses. This is because one of the problems that we've seen with LMS (learning management systems) inside organizations is that all these applications are created, but nobody uses them. Generally, the reason is that there's no buy-in, there's no adoption; so, it is very important to also drive end-user adoption and do the change management needed to create incentives to give them the basic skills that they need in order to be informed users.

To accomplish this, you need a combination of incentives driving people to use these tools and a champion or champions to help inspire people to use them. The incentives get at, why should I learn? Why should I do my job differently? Then, you need to create a champion to serve as a change agent–an AI champion. You also need to perhaps redo your compensation mechanisms or provide some sort of incentive for doing this. Then, it's crucial to give people access to these tools, platforms, and applications in a way that is easy to use. If you build a learning pathway application, but it's very hard to use, or if people don't like how it looks or how it works, they're not going to use it.

This is where user interface and user experience design become important. But there's also a larger question of upskilling for the entire employee base, which is relevant because AI will not just automate the way people work. It will also change the way they work. Ideally, it's not just about automating a flawed process. It's about redesigning the process and, in a world where AI and humans work together, the job that the human does is now different from the job that used to be done in a human-only world. Now, humans need to learn how to partner with AI. It's almost like you now think of AI as your digital co-worker, your assistant that's working with you; you're going back and forth with them saying, "Hey, help me with this, help me with that."

But you need to know how to actually work with this digital coworker. You need to be comfortable with the fact that there are parts of your job that you will no longer be doing. It's automated. It's run by the AI system. You also have to understand its weaknesses so that you are not taking everything at face value as gospel because AI systems can hallucinate; they will make mistakes, and thus human oversight is essential.

For example, going back to the marketing use case mentioned previously. In the pre-AI days, you might have been a person responsible for creating content. You're the creator. You write blog posts, you come up with videos, you do imagery, etc. You also may have a website and do social media posts. That's a pre-AI content creation role. Various parts of that role can now be automated with AI. So, your job now shifts from content creation to content strategy review and orchestration, which involves asking, "How do I place the different pieces, and is my overall creative strategy working?"

Now, you're generating thousands of pieces of creative content using these tools. And you ask, "Am I providing the right level of oversight? Is the AI following the brand guidelines? Is there any offensive language? Are the translations to different languages working correctly?" That is orchestration. It's the same thing with marketing campaigns. You're not going to be executing the campaign, doing the A/B testing yourself. You're going to be orchestrating. Orchestration is a higher-order skill. In terms of upskilling, now you have to learn how to actually manage an army of agents, or digital co-workers, that are working for you.

How do you supervise digital co-workers? How do you work alongside them? Sometimes you work alongside them, where it's a human in the loop, and you're going back and forth with the AI. Sometimes you're working as a supervisor. And sometimes, you're simply providing high-level oversight while the AI is running autonomously. So, there are different levels of human involvement and different models and modes for human AI pairing. But it's important to understand what the right human AI player mix is because, by the way, human plus AI will almost always outperform human only or AI only.

When working on upskilling, the question is, how do you perform those orchestration skills? Do you know what AI does best and what humans do best? Do you know how to make sure that the AI is not going off the deep end, that you're providing the right level of oversight, but at the same time getting the most out of the automation of the tools? If you continue to do the work manually, you're not being productive. But if you give the AI too much leeway and leave it alone, it can make mistakes. So, that is going to be a very important skill. The most egregious examples of AI going wrong are these AI chatbots that people will abuse to get them to say things like it's great to attack others, or that a certain group of people is terrible, etc. When that happens, they have to take down the chatbot so it can be reprogrammed to avoid that type of behavior.

A challenge associated with upskilling for AI is that a lot of this learning is experiential. You have to actually interact and work with the AI to learn it. You have to integrate AI tools into your daily functioning to understand them, benefit from them and guide others on how to leverage them. You should be using AI day in and day out, at which point it becomes a skill. Spending hundreds of hours with platforms such as Perplexity, ChatGPT, or Gemini will give you a good grasp of their capabilities and limitations. You will know what they do well and what they don't do well. As a result, you will be able to get a lot more out of these platforms than an individual who doesn't use them much because you have practiced–you have learned to understand the nuances.

It then can become puzzling, or even appalling, to see that some companies spend lavishly on providing access to all these tools, but nobody uses them. They don't use them, and they don't know how to use them because they have not practiced. It goes to our point above that to get good at using these tools, there is no substitute for experience. There is cognitive understanding, and there is visceral understanding. Cognitive understanding is, "I kind of know what AI does." Visceral understanding is not brainstorming; it's body storming. It's actually experiencing it. That's what you have to do. Unless you sink your teeth in, unless you put your toe in the water, unless you dive in and you experience and you understand, you won't be able to access the full power of these AI tools. You can build the world's greatest AI tools for people, sponsor the world's greatest AI initiatives, but if people aren't going to engage and sink their teeth in and work with them, they're not going to have any impact.

External Partnerships

With regard to partnerships with external technology providers, one of the interesting challenges is how well these AI tools are going to work out of the box and how much they need to be customized. How much elbow grease is needed to make them actually work within your organizational context, within your systems context, technology context, business process context, etc.? Now, if you go back to the days of enterprise software, why did we create the big four? Why did we create these giant consulting firms like Accenture? Because enterprise software is hard. When you say, "I want to do an SAP implementation," for example, it has been said that for every $1 of software SAP sells, $4 is paid to partners because it is so complex to implement something like an ERP software system.

You might end up with a $100 million project with armies of consultants running around.

If we fast forward to the world of AI, we're saying we have all these AI tools, but are they going to work right out of the box? No, they're not. You're going to need elbow grease, just like with ERP solutions. That's another interesting question for organizations planning AI projects: to what extent do you need these additional services? Alongside the products, what is the right mix of services and products of the vendor? So, when you look for a partner, you also have to look for customization, services and consulting expertise to make the AI initiatives work. That's something else to look for: what is the level of support that you're going to get, the level of solutioning you're going to get, from your partners?

That's one kind of partner, the technology partner. The second type of partner that you'll work with is your downstream partners, like your channel partners, resellers, value-added resellers, distributors, etc. These are the people who help you take your product to market. For those folks, what you have to figure out now is how you're going to actually move data across these AI tools so that you can reap the full benefits of the AI. There has to be trust; there has to be transparency with your partner. That is more an issue of how we collaborate around the data, while managing the information, security and privacy?

CHAPTER SEVEN

Execution Excellence: Turning Vision into Action

"I think there are two parts to the execution piece: one, you need to define very clear ownership. And two, you need to have very clear metrics, because you don't scale AI through pilots, you scale it through process."

Ram Iyer - General Manager, Windows Devices & Sales, Marketing, Microsoft

Great strategies often fail due to poor execution. This is why the fifth Value Lever–Execution–is vital to successful AI initiatives. This chapter focuses on how organizations can turn AI visions into tangible results through strong execution capabilities. It focuses on topics such as securing leadership buy-in, managing change effectively, AI adoption culture, and ensuring sufficient resource allocation. A case study of a voice-based insights platform illustrates how execution missteps, such as unclear scope and lack of appropriate planning, can hinder AI adoption.

Execution requires robust coordination between business teams and technical teams, as joint efforts help ensure that the right planning is done and everyone is aware of what will be needed on the ground. Sometimes, businesses tend to think that the AI will be a "magical wand" that will just come in and solve the relevant problem. The reality is that solid execution and planning of that execution are equally or more important. Throughout this chapter, readers will gain practical guidance on overcoming these challenges and building the organizational muscle required to execute AI projects with precision and impact.

Execution is where the rubber meets the road with regard to the success or failure of AI projects.

This Value Lever has four subdimensions:

- Resource availability
- Change management and the feedback loop
- Leadership support
- AI adoption culture

We cover these subdimensions in detail in the following sections:

Resource Availability

Resource availability centers on whether a program is resourced appropriately across the project life cycle. This refers not just to the resources dedicated to the functional part of building the algorithm, but also, and often more so, to resources needed for ongoing implementation and monitoring. Are the right people available? Has the deployment infrastructure been appropriately planned for?

A very low rating in this category means a project is significantly understaffed both for skill and effort, while a very high rating indicates resources are available throughout, from AI model building to implementing it throughout the organization. This involves effective planning up front to determine what is needed and ensure there are no schedule or budget delays. This goes hand in hand with the problem structuring that we talked about in Chapter Three; if a problem is well structured, it will enable better deployment planning and help organizations understand what they need and how to prepare for it.

If resources are delayed for either funding, infrastructure, or skills, it can lead to time and cost overruns, and in some cases to programs not getting adopted at all. The reason is that acquiring resources might take too long and, by then, the work done may be redundant, or may need to be redone as business dynamics have changed. In our experience, organizations often do not do a thorough enough job of budgeting the appropriate resources for these projects.

Change Management and Feedback Loop

By their nature, AI programs typically bring about change. A robust change management program is necessary to inform people what is changing and how it will impact the company and its staff. An important part of change management is identifying what that change is, structuring it and then communicating it. If you're not able to do that, the chances are people will not understand what the change is and, therefore, they will not adopt it, resulting in program failure. It is also critical to build an effective feedback loop. This could be system-driven or people-driven.

When AI models are directly embedded into a process, you can take the results from that process and feed them back to the model so it can be improved over time with more data. Moreover, business and environmental dynamics change, and models can drift and need to be updated. Organizational teams that are using the outputs of the AI project will also notice possible improvements and desirable enhancements. Capturing this feedback is critical to ensure that the AI provides outputs that are useful and in a format that is easily consumed.

A change management process that is nonexistent is the lowest level. The next level is change management planning that is not comprehensive; it exists, but does not cover the full scope of the plan. That would be rated as a 2. A score of 3 would indicate that change management planning is performed, but without appropriate internal communications or feedback loop.

A score of 4 would indicate change management execution per plan with effective communication and ad hoc closed loop feedback from business stakeholders. Very high- or 5, would mean execution as per plan with effective communication, evaluation and system-driven closed loop feedback from business stakeholders to enable learning and improvement. These ratings are driven by the insight that enacting change management with systemic feedback from business stakeholders makes it much more effective. A constant cadence of feedback enables those driving the change to adjust their approach as needed to optimize efficacy.

The Key to good change management is to get the buy-in of the people who will be involved before you launch the project. Make the business and personal case, and have them participate in the planning process. Ideally, depending on how much change you're planning, you should have within that team a few citizen AI scientists. These people can act as the bridge between the technology and the less technical members of your staff. They don't need to be deeply technical; they're just people who are a little bit more savvy, more curious, more willing to experiment and learn new things. Having some of those people supporting a project really helps. One part is getting the buy-in, the other part is about documenting the process. This includes what it was before, what the new process is going to be, and communicating with clarity to all the people involved that this is how things will change. Then, set clear timelines and measure as you go along as to how you're doing against your plan.

Change management has been demonstrated to be a critical factor driving success in terms of its interaction with other variables throughout our model. Now, with agentic AI programs being deployed, we anticipate that change management will become even more critical as organizations look to restructure processes and workflows with the right Human-AI combination. People's roles will be radically changed in these new workflows, and process re-engineering and change management will become critical here.

"After building and implementing an AI model, I believe it's important to consider the entire change management feedback loop and recognize that the intervention will inevitably lead to shifts in user behavior. Monitoring and responding to these changes is a critical part of successful deployment. Using a classic HR example: In the context of predicting turnover, the model will provide the likelihood of someone leaving and will identify the top three to four contributing factors. However, over a period of six months, factors that were previously significant may lose relevance due to changes in market dynamics, evolving Associate needs and behaviors, shifts in business operations, and competitor strategies. Therefore, it becomes essential for organizations to regularly review and update their predictive strategies to ensure continued relevance, effectiveness, and alignment with the overarching vision and business objectives."

Ujjwal Sehgal — Global Head of People Analytics, Mars

LeadershipSuppor t

Very low in leadership support means that leadership is not aligned with the project's priorities and proposed outcomes.

Typically, we've seen that this gap happens for two main reasons:

- Inadequate communication between AI groups and leadership prevents the AI teams to properly understanding the business priorities.
- AI teams are unable to communicate and explain the impact of what they are setting out to do.

For every program, it is critical to pre-define the expected impact and establish metrics by which that impact will be measured. This usually goes a long way in facilitating the right dialogue with leadership. Sometimes, having citizen AI scientists can help bring that about. A rating of 3 would be leadership that is aligned with the proposed outcome, but the priority is medium. The rating is basically determined by the combination of alignment and priority of leadership. A very high rating would mean leadership is aligned on both outcomes and priority, and views the project as of vital importance. In this case, the project is a strategic priority for the organization and will benefit from having sufficient resources and organizational focus.

Leadership support is critical because it starts at the top–that's where the guidance for implementing these initiatives needs to originate. Basically, AI brings about a change in the way of working because people have become used to doing things a certain way, using intuition or habituation, working with certain types of approaches and technology. All of a sudden, they have these models that tell them what to do; then, there are models that are doing it for them. Beyond that, we have this new type of human and AI-augmented work that is being created. This could be, for example, creative work like marketing or creating content together. It's a new way of working, and people have to adjust to that.

To accelerate and embed that adjustment, the push has to be straight from the top; the people at the top have to make it very clear that this is needed. They have to speak about why it is important. They have to evangelize it, and they have to lead by example. If they haven't bought into the initiative and aren't fully supportive, that won't happen.

To get leadership buy-in, the motivation or justification for supporting the AI project has to go hand in hand with the business case. The business case needs to be clearly evaluated, understood, and proven; ideally, the AI impact model should be used to measure the impact of the business case beforehand. Doing so enables an IT team or the project management team to demonstrate to their leadership: "This is the impact we're expecting from the program." Otherwise, they would need to create a business case and make sure that it aligns with leadership priorities. It shouldn't be AI for the sake of AI or technology for the sake of technology or doing something that is just cool and funky. It has to be something that aligns with leadership's technology priorities and vision for the organization. That is vital, and making a business case is critical to get that buy-in.

Beyond the leadership, you also have to get buy-in from the team, the people who are actually going to use this. You need the leadership's help to get the team to buy in, but not just because the leadership is pushing it. The team has to actually believe that this is going to be better–this is going to help. This especially applies to people with a lot of experience in your organization who've been doing certain things for a long time, and they have certain inherent knowledge and domain understanding. If they're going to adopt AI, they have to understand that it is going to help them do better and get better. So, make the business case to them too. These people are crucial to getting the people who are going to be touching and using the results of this program behind it.

"What I see happening over and over again with AI projects is this notion of a field of dreams. In other words, if you create a great solution, people will naturally adopt it. In reality, they won't or at least not enough adopt it to make it viable or scalable. Some might say that the answer is to mandate adoption of your AI solution, but that can be too heavy-handed and stifle innovation. You need to give people something that is motivating, useful and helps them perform their roles better and more productively than what they have today. However, you must be able to both encourage its use and have mechanisms or controls in place to support it. This is where most companies fail.

AI adoption fails because companies fall short on their willingness or ability to create an ecosystem to drive adoption. It's a balancing act between driving adoption and encouraging innovation. Companies that do this well will achieve a greater degree of success in AI adoption."

Doug Hillary – Strategic Advisor, C5i, former SVP of Analytics at Dell

AI Adoption Culture

This subdimension focuses on the culture that exists at an organization with respect to the adoption of AI and the data and analytics associated with it. The level of adoption and support of AI within the culture will play a large part in determining the success or failure of an AI project. This is one of the most important factors that came out of the model.

Is the business leveraging insights from analytical programs in supporting its decision-making? Does the business see the value in content from generative AI and feel that it can adequately support their work, and that the right Human-AI combination can lead to better outcomes?

With regard to data supporting AI analytics within a company's culture, we can use the example of forecasting to illustrate what North Star means in this context. Forecasting can be performed by, for instance, using historical data over the past two years based on shipment data of a certain product. Shipment data is not direct; the product has been shipped to meet the customer demand, but the real demand lever is actually the point of sale consumption at the retail checkout counter. The data coming from that source tells you what the consumption has been. Thus, North Star would mean to be able to use that type of data to come up with the most accurate demand forecast.

This would involve using point of sale data, using demographic data, using X, Y, and Z and other kinds of data that are important, and then coalescing that data to come up with more accurate quotes.

In general, forecasting will be more accurate if you're going to ingest a significant amount of data. But then, you need to have the capability to parse through the data and bring it to a certain level, as well as the ability to combine the data. So, if you're going to do forecasting for a product or a certain segment, let's say young folks in the Midwest of America, for a product like hair bands or something to do with apparel, you would want to understand the demographics in the Midwest region. How many males or females? What age group, 18 to 20, 20 to 25, or 25 to 30? Then, you marry it with the historical trend. Which kind of apparel features are performing better than others? That data is going to come from somewhere else, but you need to have the wherewithal to combine all of that data and make sense out of it. There has to be some sort of technology used for that purpose. That revolves around the ability to combine a substantial amount of data and then create a more and more accurate forecast.

North Star means the best possible way of doing a certain thing. In the example above, forecasting is most accurate when coming from a combination of different data sources. That's the best way to get the most accurate forecast. It is critical when the forecast comes out that the relevant people are actually going to consume it and leverage it to make key decisions as to what needs to be stocked for how much, where, when, etc. Then, some people will make further decisions as to what should be produced and in what quantity and timeline. This entire process will involve many people and many systems along the way.

Very high takes the North Star capabilities and makes them all available to everybody at a cost. Because of automation and AI capabilities, this cost is, ideally, a fraction of the benefits gained. Thus, you improve your top line, but you reduce your cost at the same time using these AI capabilities and therefore boost your profit margins. The insights that are created are available to everybody in the same form. Everyone gets the same data feed and the same view of the insights. Typically, what happens in undemocratic processes is that there are some people who will know, some people who will not know, and they will not talk to each other and therefore make decisions just on the basis of instincts or "gut feel." As a result, those decisions will often not be the right ones. The difference between high and very high is a more democratic dispersion of the insights and more democratic feedback. It goes beyond the North Star through augmented analytics and allows you to get more opinions and more feedback because you're giving those insights to everybody as opposed to just a group or silo.

Another example is where generative AI is used to create sample visuals and taglines for marketing campaigns. Do the marketing managers believe that this can give them more options, improve their productivity and lead to a more impactful campaign overall? Often, the best outcome is achieved with the right combination of human perspective and AI power, but if people are not convinced that this is the case, it will not work.

AI adoption culture is very difficult to change. It needs substantial communication of the organization's intent and the benefits anticipated. Senior management needs to lead by example here. In some cases, it may involve anticipating and communicating openly about fears of job replacement. It is also important to demonstrate quick wins to build confidence. If the organization embarks on "big bang" initiatives and tries to "boil the ocean," it can cause substantial disruption and derail the agenda.

"It's vital to celebrate successes or create an environment where we are quickly trying things and failing, and then also, consequently, when they are successful, trying to elevate that and get everybody to see that it's actually working."

Ram Iyer – General Manager, Windows Devices & Sales, Marketing, Microsoft

Collaborating with AI

At a high level, we are describing collaborating with AI as a co-worker. That is a new idea, because software has never talked back to us previously. Now, software talks back to us, and we can have a conversation with it. We can have conversations with data, we can have conversations with knowledge, we can have conversations with documents, we can have conversations with bots. You can go back and forth, and you can learn and teach each other.

Another very interesting thing about LLMs like ChatGPT is that they now come with memory. This means that every time you talk to it, you can say, "Remember what I told you." For example, if you ask ChatGPT to help you make a presentation with visuals, you can use your conversations in that process. If you have told it in the past that, for example, Northwestern University's colors are purple, and the font that you like to use is that color and that you don't like text in your images, and that you like them to be abstract, you have trained it. Now, you can just say, "Make me a picture of a product manager looking at a roadmap." You don't have to tell every single detail. It will build the image based on what you have told it in the past.

If you ask it to make you a presentation image as a cold start, the results would likely be much worse; they wouldn't adhere to brand guidelines. The more that these systems learn about you, the more personalized, the more accurate, and the more meaningful the output is.

But again, how do you get the AI to do this for you? The key, as we have stressed, is that you have to work with it–experience it. In this example, the memory that it has built up about you is much deeper because of the time you've spent working with the AI. It's almost like a relationship, in some sense. For example, your significant other gets to know you well because they know your quirks, they know what you like to watch on TV, the restaurants you like to eat at, all that personal data which people would likely not know just from your public persona. We can think of a couple that has been married for twenty years as an example. They instinctively know not just their spouse's habits, but they can often sense when the person is having a bad day or their moods, etc. So, you can almost consider these AI bots as kind of an alter ego and interact with them that way.

It should be mentioned that this is the world of generative AI and agentic AI. This is not the world of traditional AI, like in the old days, where what you would do is run a segmentation model and say, "Find me the customer that I should prioritize." The model would then run it and come up with some data. That's not conversation, that's just using models at the back end. Now, it's a conversation we have evolved together. We learn together, and as you make the tool better, the tool makes you better–it's a collaborative effort. That is a new skill.

In this new world that AI is ushering in, collaboration across people continues to be important, but imagine that you have ten members of a team, where six are people and four are bots. Now, collaboration needs to happen not only between people but between people and AI, and AI and AI. It's almost like we're going to create these hybrid workforces or hybrid teams consisting of digital and human workers. Now, collaboration becomes, at least to some degree, an idea that transcends human beings. So, we take the upskilling needed at the individual level with respect to using AI, but now the same applies at the team level too. It's like we are together collaborating with a set of AI tools to create a better output. There's the idea that we had at the individual level taken to the team level.

Another dimension of collaboration is that if you're going to build these AI tools, you need these different skill sets, and they need to be able to work together. Another collaboration challenge that arises is that organizations are not designed around these new initiatives or new tools. Thus, you have to work across organizational silos to implement them. The natural structure of an organization is often not well aligned with the natural structure of these AI initiatives. So, you now have to figure out how to actually work this way as opposed to working in silos. And the silos are not going away in the short run. That raises the question: how do you, in your day job, still report to your manager in a silo but also work on these hybrid human/AI teams and collaborate across functions? That's a challenge people will increasingly need to deal with as AI gains in adoption because we don't have these lateral structures as well defined as the vertical structures.

Challenges to Execution Excellence

Unclear Scope

Unclear scope is related to problem definition, which we discussed in an earlier chapter. There is a high level of correlation between the two–the better you define the problem, the more likely you are to avoid running into issues with unclear scope. If a problem is well structured and the scope is clear, it is easier to plan effective execution. It is not enough on its own; you still need to execute effectively, but it gives you a solid base–a platform–to build from.

If your scope is not well defined, when it is time to execute, you will generally not have planned things out well enough; in such a case, you will have to really scramble to try and make the project work, with no guarantee of success. Maybe you'll get lucky, but often it will not work because of a lack of advanced planning. This typically means that you did not involve the right people, align the right resources, nor set up the right systems. If you had done your problem structuring well, the scope would have been clear, and you could have planned your execution well. If the scope of a project isn't clear, the likelihood of failure rises substantially. You could still succeed, but to do so, you'll have to be very dynamic and agile.

While properly defining the scope of the problem in the first place is the best way to avoid encountering unclear scope, sometimes the results of the program you are developing can help you narrow and restrict the scope. Sometimes, you may not have identified exactly what you hope to achieve. Do you want to increase profit? Do you want to change the product mix? Do you want to raise prices? Do you want to reduce costs? Do you want to cut inventory? Do you want to reduce turnaround time in marketing campaigns? After doing some exploratory work, the data analytics you generate can give you some direction. This can narrow the scope in which you wish to operate. But again, it may be too time-consuming or too expensive to make such changes once you have launched a program.

The ideal situation is to have a well-defined scope up front, of course, but if that isn't the case, you can use certain insights you have gained in launching the project to quickly realize your scope and really focus on making it happen; then, the challenge is that everything needs planning and preparation. If you're going to, for example, change your product mix, it likely won't be easy. There would be much you need to do to make that a reality in terms of product planning, design planning, production and procurement, etc. To reiterate: having an unclear scope may work at times, if you are very nimble and dynamic. However, this is generally not the case, which is why it's preferable to clearly scope a problem up front.

Resource Allocation

Ensuring sufficient resource allocation to get the project done centers on advanced planning. Once you get the leadership buy-in, you need to know what resources the project requires. Getting resources allocated is one thing. But it starts with figuring out what you really need. Depending on the initiative, you may need, let's say, a certain amount of money. If you are going to spend on campaigns, you may need access to certain technology, like orchestration technology or a system of insights. You have to figure out what technology you need, as you may need certain permissions within your organization from your tech team or other teams to use certain platforms; if you want to send emails to certain customers, you may need permissions for that.

First, figure out what resources you are going to need. Then, in the process of getting leadership buy-in, it's essential to communicate to leadership right up front: "These are the resources that we need." Some of this may change because resource requirements can be dynamic depending on the output of your program. Therefore, it's important to communicate that, for example, "This is what we estimate, and this is the variation range depending on the results that we get."

Case Study: Business Insights Chatbot

Company: Global Technology company

Objective: Provide automated business insights

Issues: Unclear scope

The insights chatbot voice assistant represented an early example of chatbot type AI, but lacking the capabilities of the LLM models popularized by ChatGPT.

This ambitious project which was implemented in 2017 ran into issues with execution challenges such as unclear scope and limited ability to scale insights, which resulted in disappointing performance. While the idea underlying the insights chatbot was a good one, due to technological limitations, the time for its deployment was not right.

With regard to the scope of the digital assistant's abilities, some business users were disappointed because it was unclear what the chatbot could do for them or in certain situations, the chatbot wasn't able to comprehend the total scope of the question effectively. Users got upset because they didn't think it was doing what it was supposed to do. For example, they would ask the assistant questions, and they would get an answer; then, they would expect that the system would store and remember answers given previously. For instance, let's say I asked the chatbot, "Please let me know if any location in the Midwest region has low inventory?" If the answer is Columbus, Ohio, the next question might be, "Why is it low? What is causing inventory to be low in that location?" This involves nesting logic, and while this nesting capability is available to chatbots now, it was not mature at that time. There were limitations on how far the system could go in giving analytical and cognitive answers, and this was not clearly understood by the business users, so there was an expectation mismatch.

To sum it up, people thought the scope of the insights chatbot initiative was far more than what it was, and this resulted in unhappy users. The takeaway is that you need to define the problem your initiative is solving for clearly in terms of its business objectives, constraints, impact, etc. If that is not done, it significantly reduces the chances of project success.

In this case, the failure to define the constraints prevented the project from delivering the functionality customers desired. That led to the unclear scope which caused the project to experience significant difficulties. In our analysis of this program, while we identified that data availability was high, we also assessed that execution of the program, in other words, the scope of the questions, was not well defined. The chatbot had a lot of data stored, but it couldn't always get you the right answer when you asked a question, and this ultimately led to its failure.

Principles of Execution Excellence

Achieving excellence in the execution phase of a project goes back to the importance of problem framing. A simple analogy is to say that the fastest person in the world, or in the stadium, will nevertheless lose the race if they run the wrong way. Execution is not just about speed; it's also about speed with direction. And if you don't have the North Star clearly defined, people can help, but they don't know where to go. This is why scoping and defining the goals and objectives is so important.

Here are some principles for excellence in execution from our experience:

One is the importance of having, as we have discussed, clarity around goals and objectives, so that everyone knows what the true North is and everyone knows where we need to go. If this is not the case, we won't get coordination, we won't get alignment, we won't get everybody pulling together. To use the analogy of a long boat, ideally, we want to have many rowers rowing in the same direction to maximize speed.

"I think that understanding where and how to implement AI is a critical factor for successful adoption. Adoption is not just about using AI but about integrating its output into every decision that you make. This element can become a significant barrier, because the final implementation phase is often the most challenging and crucial to achieve. For example, I might get a fantastic output, but if I don't know how to use it, that output is of no value to me."

Ujjwal Sehgal – Global Head of People Analytics, Mars

Also related to execution excellence is clarity around who does what roles and responsibilities. If we don't know who's supposed to do what, who is responsible, who's accountable, who has the authority, things get slowed down because we're just looking at each other, saying, "Well, you're supposed to do this, you're supposed to do that." In such a case, execution suffers. Also, business leaders not only need to clarify what the goals are and to paint a clear picture of what that entails, but they also need to inspire. They need to motivate people by saying, "Now that you know what the North Star is, now that you know what the destination is, mobilize yourself, get things in motion." That requires constant communication and a sense of accountability.

If you want efficient execution, as the saying goes, *inspect what you expect*. People need to know that they're going to be held accountable, that the boss is watching, that these are your metrics, these are your goals. So, after having defined the KPIs, the goals, OKRs, whatever you want to call them, you need to create a culture that says, "We have to deliver. and we're serious about meeting the numbers and the deadlines and these goals." That is an additional aspect of execution excellence, and it helps explain why lack of clarity around goals, lack of alignment, and lack of inspiration and motivation can all combine to get in the way of effective execution.

"Where I think AI projects often break down in the execution phase is the unwillingness or inability to create an environment for success. It isn't just about having a great idea to solve a problem or encouraging innovation and experimentation with AI. You need to have all these other mechanisms in place and judiciously applied so that you can allow ideation, but then control the use of your resources, and direct those resources to adopt and utilize AI in a way that adds the greatest value for the enterprise."

Doug Hillary – Strategic Advisor, C5i, former SVP of Analytics at Dell

When evaluating strategies underlying execution, it's important to focus on the tangibles; the challenge with execution is that it is the glue that connects strategy to operations to people. It's a connective tissue; it's really where the rubber meets the road. For the rubber to meet the road, to get execution right, you need to get a lot of ducks in a row: strategy, operations people, tech, data, etc. Everything needs to come together in the right way. An analogy that can be helpful in this regard is to think of strategy, technology, data, and all of these things as being like potential energy.

Execution converts potential energy into kinetic energy. It makes things happen. So, it's a big waste if you've created a lot of potential through great data, great talent, technology and so on, without being able to effectively utilize it. If you are not able to mobilize it, if you're not able to convert it into action, then the whole thing is a waste. And that's the power of execution. As one of the authors (Dr. Sawhney) likes to say, "Strategy is ten percent strategy and ninety percent execution."

One idea for optimizing an organization's ability to execute AI projects effectively is to set up AI squads. These are cross-functional teams that are given specific initiatives. These squads consist of four types of capabilities:

1. People who understand the domain of the business problem or the use case.
2. Data scientists and AI experts who understand models and data.
3. People with IT expertise who understand your enterprise system to enable it to talk to the system of records.
4. Product management capability, somebody who manages the project as a product manager.

Having all four of these capabilities is critical for rapidly executing on a project. This is especially true in larger organizations where, sometimes, people really spin their wheels in terms of: who is responsible for what? And how are we going to do this?

Another problem big organizations can face is that silos often don't talk to each other. That is why it is vital to establish these cross-functional squads to bring people together. There is also a fifth kind of expertise that is sometimes necessary on these AI projects, especially for organizations in a regulated industry, and that is compliance. When substantial compliance issues are involved, it's good to bring compliance expertise onto the squad, to bring in some of the legal experts because they're the ones who will have to say yes or no at the end.

By creating these squads, organizations can speed up execution. This is also where the idea of agile processes comes into play. These squads can generally benefit from using agile methodologies. They need to iterate fast. Thus, in large organizations that have a bureaucracy that appears to be highly siloed, creating these cross-functional agile teams can prove valuable. You want them to execute objectives against specific tasks, so you might have squads in HR or squads inside marketing or inside sales working on AI use cases related to those functions.

All of these squads and their activities can be placed under a single AI Center of Excellence. Essentially, this enables organizations to drive execution forward under a unified source of command. That AI Center of Excellence is a convening body. It's a hub that coordinates the spokes that are made up of these squads. It helps provide them with the direction, resources, and authority that they need.

CHAPTER EIGHT

Interplay among Factors: The Role of Interactions

"AI ambition consistently outpaces readiness, given the transformative potential of the technology. This ever-widening gap can be narrowed down to three key considerations: infrastructure, context, and trust. Firstly, secure and efficient token generation is the new competitive metric, but infrastructure capacity and fragility around power, compute, and network connectivity are real constraints that don't get enough attention. Additionally, without context to rely on, agent decisions are guesswork at best. Bridging this divide means connecting enterprise data, enriching agents with machine data, and embedding AI into newer, redesigned workflows. And finally, the AI risk today is no longer a wrong answer - it's a wrong action. Safety and governance must be a part of the blueprint from the start, not bolted on later. The foundations of the AI-powered future are being laid right now, and it's our collective responsibility to ensure they are built to last."

Daisy Chittilapilly – President, Cisco India & South Asia

The value levers in AIIM do not work in isolation but are deeply connected. Sub-factors within any value lever could be connected to sub-factors in the same or other levers. An interesting example shown by our model is where "Change Management and Feedback Loop" and "AI Adoption Culture" were not in themselves significant determinants of success, but influenced the impact on success of various other sub-factors. Similarly, our model threw up insights into tradeoff variables, showing which factors should be prioritized over others.

One key tradeoff variable showed that optimizing data is more valuable than optimizing tech.

The role of each factor is nuanced depending on the nature of the program and the other factors involved. This chapter will help readers to understand these inter-relationships and simplify the complexity in planning for the influence of these factors on the outcome of AI programs. By applying the AI Impact Model to understand these inter-relationships and tradeoffs, organizations can make informed, strategic decisions that maximize the return on their AI and analytics programs while navigating the constraints and complexities of real-world implementation.

The results of our analysis suggest that successful implementation of AI initiatives requires a multifaceted approach that addresses both organizational readiness and technical capability. Clarity and strong support from leadership are foundational, ensuring a unified vision and commitment to integrating AI across the organization. This must be complemented by a well-defined change management process that steers the organization through transitions, mitigates resistance and aligns teams with new workflows and technol-ogies. Also critical are well-structured problem statements and clearly deployable use cases that center on high-impact opportunities, enabling AI solutions to effectively address specific business challenges. Highly recurring AI use cases usually drive better results.

A robust AI adoption culture is essential, where decision intelligence is democratized through augmentation and automation, empowering employees at all levels to leverage AI-driven insights in their decision-making. Central to this is an insights platform that ensures insights are not siloed but distributed seamlessly to the relevant stakeholders, fostering collaboration and more informed decisions. Additionally, the availability of high-quality data, coupled with a growing pool of citizen AI scientists, is vital for the widespread adoption of AI solutions. Together, these elements build a strong foundation for organizations to harness the transformative power of AI and drive meaningful business outcomes.

AI Project Selection and Sports

In Chapter One, we compared the process of selecting AI projects to that of a professional sports franchise choosing which players to draft or acquire via trade. In this section, we'll compare how the various Value Levers and subdimensions work together to determine AI project success to a similar process in the world of sports.

As discussed in Chapter One, simply looking at a player's statistics is generally not enough for a professional sports team to determine if that player is likely to be successful at the next level. This is because there are a variety of factors that work together that must be considered. Is the player going to sync with the culture of the team? Can they adjust their approach as needed? Do they have the physical skills to be successful at the next level for the specific role they are expected to perform?

Similarly, when gauging the potential of AI projects, enterprises must determine how a variety of factors related to a project are likely to work together to enable it to succeed. A project may appear to have great data, the technology used to orchestrate and generate insights may be top-notch, the talent may be there, and yet the project may still end up failing.

In the same way, simply filling a team with the greatest stars does not necessarily mean that the team will win. The reason is that there are various other factors at play that drive the end outcome–some have to do with the team, and some have to do with the dynamics on the day of the match. For instance, players may be unable to overcome clashing personalities which leads to resentment and subpar performance as a team, even if, individually, the players possess tremendous talent. Additionally, if the team hasn't practiced together effectively and they meet a less skilled team that has worked harder in practice and has a better game plan, their superior talent may prove insufficient to secure victory on the day of the contest against the less talented but more prepared opposing team.

The following is an analogy of how our five Value Levers might be used to demonstrate how various factors play into a team's success or failure in a sporting event.

Problem - This has to do with the characteristics of the team members themselves and whether their skills are complementary, allowing them to achieve synergy when working together. It also applies to the quality of the opposition the team faces–the better the opponent, the more important it is to work together well if the team hopes to be victorious.

Data is the foundation - This relates to the quality of the training the team has received and to their nutrition, sleep and mindset before a match.

Technology - The ground and weather conditions, and whether the team is playing on their home ground or away. also, the quality of their equipment.

Talent - This would not only be the talent of the team itself but also the supporting talent in terms of the coach, physio, and other support staff.

Execution - This refers to cultural factors such as team spirit, motivation, support from team ownership and management, as well as fan support.

Some of the most famous victories in sporting events over the years have involved teams widely considered to be underdogs overcoming seemingly tremendous odds to achieve victory against apparently superior opponents. The *Miracle on Ice* victory of the plucky US hockey team against the mighty USSR team in the 1980 Winter Olympics is one example. In that contest, the seasoned professionals of the Soviet team were expected to steamroll the little-known players making up the US squad. However, the US team's passion and superior teamwork on the day enabled it to score a historic victory over the stunned Soviet squad.

In International Cricket, one of the greatest upsets of all time was India winning the World cup in 1983 by defeating the mighty West Indies squad in the final. India was not just underdogs–they were widely expected to exit early. Before 1983, they had only won one World Cup match, in 1975. Yet they beat several top teams to reach the final and defeated the two-time defending champions the West Indies, in that match.

The reasons for the victory were a rare convergence of leadership, conditions, mindset, and team collaboration. Kapil Dev's leadership changed India's psychology. He instituted an aggressive mindset, a bowling-first mentality (they stopped seeing themselves as a batting-only side) and some sharp tactics (they didn't try to overpower the West Indies but to out-think them). This result demonstrates the value of leadership, teamwork, and complementary skills, which led the Indian side to victory over a team studded with global superstars.

Another notable upset occurred in professional basketball when the Boston Celtics defeated the star-studded Los Angeles Lakers team in the 1969 NBA finals. This upset was perhaps not as surprising as the *Miracle on Ice* and India's World cup win but nevertheless showed that having a team built around star players does not necessarily guarantee victory. The Lakers, led by future Hall of Famers Jerry West, Elgin Baylor and legendary center Wilt Chamberlain, were heavily favored to win the series as the Celtics, despite having won numerous championships in that decade, were widely considered to be old and past their prime. Nonetheless, the Celtics defeated the Lakers in Game 7 of the finals. The Celtics used their superior teamwork to overcome the Lakers' advantage in star players. The result demonstrated that simply fielding a team of superstars was not enough to succeed at the highest levels of professional basketball.

In European football (soccer in the US), the victory of the German side over Brazil in the 2014 World Cup semifinals by the shocking score of 7-1 was described by The New York Times as "utterly unexpected, historic and inexplicable."[8] As with the Celtics victory, the fact that Germany won was perhaps not a tremendous surprise, given the success Germany has had over the years in team competition at the world level and injuries to two key Brazilian players before the match. But the magnitude of the victory could be said to be an example of a team that had fewer "superstars" using teamwork to overcome the individual virtuosity of the members of the opposing team. The Brazilians, despite their offensive prowess, found that their team defense was not sufficient to counteract the German team's cohesiveness and precision on that day.

The thesis of this chapter is not, of course, to say that successfully deploying AI projects draws on the same specific criteria it takes to build a successful sports franchise. Instead, the contention is that similar dynamics exist at a macro level between the two, and looking at the analogous process in sports is a helpful way to provide insight into how factors interact when planning and implementing AI projects. The conclusion to be drawn, from the authors' perspective, is that while it is important to put together all of the elements of a well-structured project, it is also vital to pay attention to the interactions between these elements–much like a sports team needs to focus on how its players' skills complement each other and help the team as a whole–to improve your chances of successfully deploying AI.

In the following sections, we explore some of the most important factors involved in Value Levers and variable interactions.

Change Management and Feedback Loop

There are certain variables that we find interact with each other and have a joint effect on the end outcome. As covered in an earlier chapter, an interesting case of this phenomenon that has a significant impact is change management and the corresponding feedback loop. What our model has shown us is that this turned out to be vital. The bottom line is that change management turned out not to have a significant business impact on its own. That was really surprising to us; it caused us to wonder if the model had gone wrong somewhere. But later on, when we used compound variables where we created combinations of multiple factors, we found that it actually was very important, as we'll describe below. We had eighteen variables. We created multiple combinations of any two or three together, and we created hundreds of different combinations using sophisticated AI algorithms to see which ones had an impact on the business result.

We found, interestingly, that even though change management and feedback loop by itself didn't seem to be significant, it had a significant effect when interacting with thirteen other variables. What it demonstrated was that while change management was not making a difference alone, for other variables to influence business impact, you need good change management. If you think of it intuitively, it makes sense. If you have great orchestration technology but you don't manage the change well, how will you use that technology? Or, if you have great data quality, but you don't manage the change well, what is the point of the great data? It is actually a compound variable that has an influence on many other variables. Thus, it's highly important as a co-factor. Overall, it may be one of the most important elements of successfully implementing a proj-ect, but it plays its role in a very different way, which is an inter-esting output of the model.

Change management is necessary but not sufficient for success. It's important to address these changes, but your project is not going to be successful unless the other factors are there as well. So, in this part of the discussion, it's important to emphasize the value of taking an ecological or systemic view. It is helpful to think holistically about what drives success; yes, there are levers, and yes, individual factors contribute. But it's also critical to recognize the importance of interaction effects among the factors in the model.

"When we talk about change management and AI, in many cases, people think that technology is very important, and obviously, it is very important. Sometimes, however, I think people just lead with technology as opposed to leading with change management, because truly, when you try to apply it without proper change management, it doesn't translate. Ultimately, we get beholden to the technology, rather than the other way around."

Ajit Sivadasan – Ex VP and GM,
Global eCommerce, Lenovo

AI Adoption Culture

Another factor where interactions are important is AI adoption culture, as we call it. Culture is not necessarily something written; it is just how your team is open to embracing new ways of working, new methods, new techniques, and changing and doing things for the better by doing them differently. As with change management, culture by itself doesn't drive success. Because if other things don't work, then culture is not going to make much of a difference. But culture actually has a huge impact on the end result because it interacts with other variables. Again, you could structure your problem well, but if you don't have a culture that is supportive of adopting and implementing the solution, then you will likely not be able to see it through or achieve the desired results from it.

If the culture is resistant, there will be internal resistance. You can build great technology, but if your people are not aligned, it will create significant difficulties when it comes to adoption. Even if they say "Yes," if they don't truly believe in it, if they don't have the mindset of, "OK, let's adopt it," or they're afraid that it will take away their jobs, it won't go well. Similarly, if the benefits aren't communicated properly, if they don't understand them, the likelihood is that it just won't work. We are all creatures of habit, and we go back to our comfortable ways of doing things when faced with change we don't understand.

"People and companies want to be AI first, and one of the things that I keep telling Associates in my organization is that we need to be AI forward, not AI first. I'm comfortable if we're second, as long as we're not getting left behind, and we're doing this correctly and diligently. We're not in a race where we must come in first place. And not in AI, especially in HR. Our priority is to be extremely vigilant with sensitive data, ensuring zero compromise or leakage because our Associates data must always be protected."

Ujjwal Sehgal – Global Head of People Analytics, Mars

Optimizing Data over Technology

Recognizing the value of optimizing data ahead of technology also came from our study of compound variables. We said, "Let's look at not just variables that have a compound effect but variables that represent tradeoffs." We asked the model to generate tradeoff variables and to help us see where investments should be prioritized if one had an option to invest between two different value levers. Data and technology appeared as tradeoff variables where, if you have to make a tradeoff between the two, then you should definitely invest more in data. Both are important; there's no doubt about having good technology for your insights platform and orchestration technology is very important, but the starting point is the data.

If you don't have good, high-quality, reliable data, then whatever technology you use is not going to matter. If you have high-quality data and you do good work, even if you don't have the best technology, while you may not get the same level of impact that you would with better technology, at least you'll have created something that's reliable. Maybe it will not be fully usable, but if it is partially usable, it will still be likely to give you results. But if you have something that you've built which is based on very poor-quality data, then you're not going to get anywhere, whatever technology you might have.

Ideally, you want to invest in both. But if you have to make a tradeoff, then data should be the focus. But then again, you have to keep in mind that, as we discussed in the chapter on data, it can have a diminishing utility curve. If you're already a 4 on data and you're at 1 on technology, you shouldn't try to take the data to 5; instead, you should bring the technology up to a 2 or higher. Please note that there is a degree of subjectivity involved in any such decision. This is not a perfect rule; it is a general conclusion. Ultimately, the right choice will depend on the situation.

Scenario: Change Management Impacts Program Success in a Surprising Way

Industry: Industrial manufacturing

Objective: Optimize operations with AI processes

Issue: Employee resistance to new processes

A manufacturer of diesel engines learned the importance of change management and feedback loops when debuting an AI program. This company was implementing a supply chain system that involved looking at the company's entire demand forecasting process, combined with their production planning. Forecasting improved accuracy, and production made it more aligned to the constraints that actually existed. The objective was to create improved production plans and schedules.

With regard to adopting the system, the company faced resistance from material planners. These planners, particularly, didn't want to adopt the system because of the perceived inaccuracies it was producing. They kept the coming up with reasons why the program was not satisfactory. The members of the program implementation team were scratching their heads as they tried to figure out why there was so much resistance to the new system.

There was one person who was leading the resistance to the AI program, and the implementation team sat down with this individual, along with a change management expert. The goal was to try to understand what was causing the opposition to the program. As it turned out, it was not an AI problem at all. An unstated problem was likely the worry among this person and others that AI would take their jobs. But the reason he gave for opposing the program was that they used to read the reports every morning from which they were ordering materials and the parts in the form of what was called a blue report. The name stemmed from the fact that the pages of the report were blue. He said, "I need to get my blue report, and I'm not getting it the way that it should be."

Based on this feedback, the implementation team made the report a blue report by printing it on blue paper. After that, the problem went away. The feeling of "change" and also the corresponding fear of job loss were substantially reduced. It was as simple, or ludicrous, as that. It demonstrates the importance of change management, even if only as an indirect factor. While change management had nothing to do with the quality of the system itself, it combined with the various factors that made the system effective to ensure that it was also implemented successfully.

The takeaway is that while the AI was effective on its own, the team implementing this new system had to do sufficient due diligence to realize that it was a non-AI issue that was preventing the system from operating effectively. The lesson is that, when implementing an AI program, it is not enough to simply focus on the technology and data side of things; the human element is just as important, and if it is neglected or misunderstood, it can torpedo an AI program even if its functionality is solid.

Further Thoughts on Connections Between Change Management and Feedback Loop, AI Adoption Culture and Other Factors

To elaborate further on the topic of change management and feedback loop, while this factor on its own did not score highly, as discussed above, it interacted with thirteen different combinations. So, a slight change in this factor will impact thirteen other variables, and those variables combined will have a far greater impact. It is about change management & feedback loop combined with AI adoption culture, the cost, the latency, the insights platform, and so on and so forth. One minor change here and there in change management impacts various compound variables that have a substantial impact on program success.

An example is demand forecasting, where you are forecasting demand and then making sure that the people who are going to consume that forecast have been trained and imparted the technical expertise and the wherewithal to understand and act upon the conclusions drawn from the data. Once you do that, the likelihood of generating the wrong sales actions declines substantially. You can also look at how demand impacts other functions like safety stock and various other stock-keeping elements; people need training to understand how to correlate across those elements.

As mentioned previously, we also found that AI adoption culture is not a variable that drives success directly, but instead interacts with everything else. Culture interacts with execution, and culture also interacts with skills to drive success. We found that the softer factor,s like culture and leadership and change management & feedback loop, play a role that can be thought of as a substrate. They affect a project in many ways, but they don't directly explain success. In regression models or in any kind of model, there are main effects which are direct effects and interaction effects, which are indirect effects. Change management and AI adoption culture work indirectly.

When you're looking at strategy, for transformation or whatever objective a company establishes, this brings to mind the old saying, *culture eats strategy for breakfast, lunch, and dinner*. What that means is that it's as if culture is the soil upon which you plant seeds. It's the infrastructure that creates the environment that allows flowers to bloom. But culture is neither the plant nor the flower. Culture is like oxygen. It is everywhere, but it is invisible. It is ubiquitous, but you can't see it directly. And it affects everything, just like oxygen affects every living being.

Similarly, culture affects every aspect of what you do in the organization because it provides the invisible yet pervasive context within which people act. Therefore, as you start thinking about implementing AI projects, it's important to realize that, in a sense, culture should not show up as one thing, but it should show up everywhere. We identified specific instances where it does show up, but the inference here is that it could be said to apply throughout the process of project implementation, even if it is not always measurable.

How does something spread its impact so widely throughout the model? It does so by interacting with other factors, by not being a standalone factor. To return to the analogy mentioned above, it's not the plant, it's the soil; it's not the flower, it's the substrate. That is why softer factors like culture, and also change management and feedback loop, are not variables that can be isolated in and of themselves. They operate on and through other factors. For instance, change management operates on people. Change management operates on technology. The pattern we observe is that these softer variables, softer as in less tangible and more interactive, express themselves through their interplay with other variables.

The two observations here are 1) the effect is indirect, and 2) the effect is secular in that it affects most other things. So, culture and change management are interacting with everything else, but they are not acting alone.

The general pattern here is that anything that deals with leadership and human beings, such as culture and change management, will tend to affect all of the other variables. In fact, when one of the authors (Dr. Sawhney) built his innovation radar years ago, he talked about different types of innovation. This included product innovation, experience innovation, business model innovation, etc. With regard to culture and change management, he found that while they didn't directly cause product innovation, they instead caused product innovation to be better.

It is reminiscent of the ads put out by BASF, the chemical company, which said something to the effect of, "We don't make Product xx (whatever it might be), but we make xx better." The things that go into the interplay of these factors involve a similar dynamic. Culture and change management don't operate by themselves, but they affect everything else.

Leadership, Culture, and Change Management

The interaction between leadership, a company's culture, and change management is an important combination of factors with significant influence on the outcome of AI projects. With regard to making a company's culture more AI adoption-friendly, or change management-friendly, leadership is the common underlying driver. Leadership sets the tone. Leadership defines the culture, and leadership is also the engine behind change management. All roads ultimately lead to the leadership of the company. This brings up interesting questions: how is culture created? How is culture perpetuated? How is culture changed? We believe it's all about the leadership.

Sometimes, when a leadership change occurs, it can help the company take the steps necessary to successfully change the company culture. For example, take a company like Microsoft under the leadership of Satya Nadella. When he took the helm, Microsoft didn't change in terms of the products or the businesses it was in or any of those tangible things, but something very fundamental changed about the culture. There was an increased level of empathy, customer centricity, and humility. Nadella had a very famous quote early on when he joined. He said we want to evolve from a company of know-it-alls to a company of learn-it-alls.

This gets at the concept of beginner's mind or beginner mindset, which, in a business context, emphasizes the importance of a company listening to its customers. The subtle cultural shift Nadella brought about led to a transformation of the company. But again, as we have pointed out with regard to interaction effects, can you draw a straight line between cultural change and Microsoft's initiatives in the cloud? No. But if you investigate the company's trajectory under his stewardship, you can see how his influence enabled Microsoft to evolve its business model successfully.

In the past, what worked for it was Office, Windows and desktop software. Then, signals arrived indicating that open source, mobile devices, and cloud were coming to the fore, and Microsoft was initially lagging in these areas. However, under Nadella's leadership, that changed, and they became better able to understand the threats and the opportunities and reposition the company. That led to better products, better revenue and a transformation of the company's business model.

It's almost like going on an archaeological excavation to find the root cause. In this case, we see a cultural shift brought about by top management. So, this is how leadership drives culture, drives change management, which in turn has an impact on products, revenue, profitability, etc. This applies to enterprises seeking to successfully implement AI as well.

To drive AI adoption, in our experience, leaders need to model the behavior. But they can't say, "You change your operational routines by using AI, but I'm still going to do things the old way." They have to embrace these tools themselves. They have to actually model their own behaviors. They have to put their money–their behavior–where their mouth is, so to speak. They have to become champions of the technologies and the tools and start using the tools themselves if they expect others to do so.

Another thing is that there has to be a forcing function for people to change, but that forcing function has to be a combination of carrots and sticks. The carrots are incentives and rewards for employees who innovate and who apply AI in fruitful ways, while the stick is that if you don't do it, there are consequences, up to and including termination. That type of atmosphere consists of a combination of three things. One is to set a personal example to give people the incentives; two is to provide them with the tools and the training they need; and three is to hold them accountable for putting those tools to work.

CHAPTER NINE

Optimizing Business Impact: The AIIM Simulation Tool

"There exists a gap between the promise of AI and the value directly attributed to it. Ashwin and Prof. Sawhney masterfully blend critical success drivers for AI initiatives into the AIIM framework. Simulating potential outcomes built on pragmatic, real-world evidence and optimizing underlying drivers over the initiative's lifespan is the secret sauce to unlock scaled AI-led transformation."

Manik Gupta — Former North America Chief Analytics and Insights Officer, Bayer Consumer Health

The AIIM framework is not just a diagnostic tool; it is also prescriptive. This chapter introduces the AIIM simulation tool, which allows organizations to predict project success and optimize outcomes before making significant investments. By simulating different scenarios, users can identify the most impactful interventions and allocate resources effectively. The simulator allows organizations to model a variety of approaches to optimizing an AI project, such as improving data quality or reallocating resources, thereby enabling data-driven decisions.

This chapter offers step-by-step guidance for utilizing the simulation tool, which has been employed by a variety of organizations to improve ROI through scenario planning. The intent is to offer practical tips designed to help integrate the tool into organizational decision-making processes.

The tool's ability to refine decisions dynamically makes it an essential component of any AI strategy. We are committed to making ongoing improvements to the model so that it continues to evolve as an indispensable tool for unlocking the transformative potential of AI.

The Simulator

We have developed a simulation tool to optimize project success by changing controllable features of a project. The simulation tool consists of three components–a ***Success Predictor,*** a ***Budget Planner,*** and a ***Budget Optimizer.*** These tools enable users to gauge a project's likelihood of success and simulate various scenarios and strategically allocate resources for maximum impact and return on investment.

The simulator offers detailed performance tracking to refine strategies (See **Exhibits 7, 8, and 9** below). It evaluates key performance indicators (KPIs), including cost-to-success ratios, overall success improvements compared to baseline scenarios, and success ratings across different strategies. Leaders can analyze these insights to compare scenarios, assess trade-offs, and dynamically adjust their strategies. By using this simulation tool, leadership teams can allocate resources, prioritize investments, and drive impactful outcomes with greater precision and clarity. The integration of the AI Impact Model ensures that every decision aligns with data-driven insights and strategic priorities, helping to maximize the potential of analytics and AI programs.

Success Predictor

The Success Predictor function of the tool can be used to identify an enterprise's most promising projects. It guides you to enter data for eighteen variables to get a success rating for the overall project. This is a score on a five-point scale that serves as an estimate of the likelihood of project success.

Exhibit 7: Predicting the Likely Success of an AI Project

Attributes	Base Score
PROBLEM	
Recurring	3
Structured	3
Deployability	4
Interconnections	3
DATA	
Availability	4
Quality	3
Cost	4
Latency	3
Risks-Ethics	4
Risks-Privacy Regulation	3
TECH	
Orchestration Tech	4
Insights Platform	3
TALENT	
Citizen AI Scientists	4
Technical Expertise	4
EXECUTION	
Resource Availability	3
Change Management with Feedback Loop	4
Leadership Support	4
AI Adoption Culture	2

Success Predictor

Success Predictor Outcome

2.94
Success Score

5 **Very High** (business impact exceeded)
1 **Very Low** (low business impact, with budget & time overruns)
4 **High** (business impact met within budget / time)
2 **Low** (partial business impact, with budget / time overruns)
3 **Medium** (business impact achieved with budget / time overruns)

Budget Planner

The Budget Planner empowers business leaders to create detailed financial plans aligned with their strategic goals. It enables users to allocate budgets across multiple dimensions, including data procurement and management, resource planning, infrastructure setup, and enhancements to orchestration platforms. The planner evaluates key variables, trends, and their impact on success, providing a clear roadmap for effective resource distribution. Users can test different scenarios by adjusting parameters such as spending amounts, controllable and uncontrollable factors, and priority items. This feature enables leaders to compare outcomes based on their inputs, anticipate the potential impact of various strategies, and make informed decisions before finalizing resource allocation.

Exhibit 8: Budget Planner

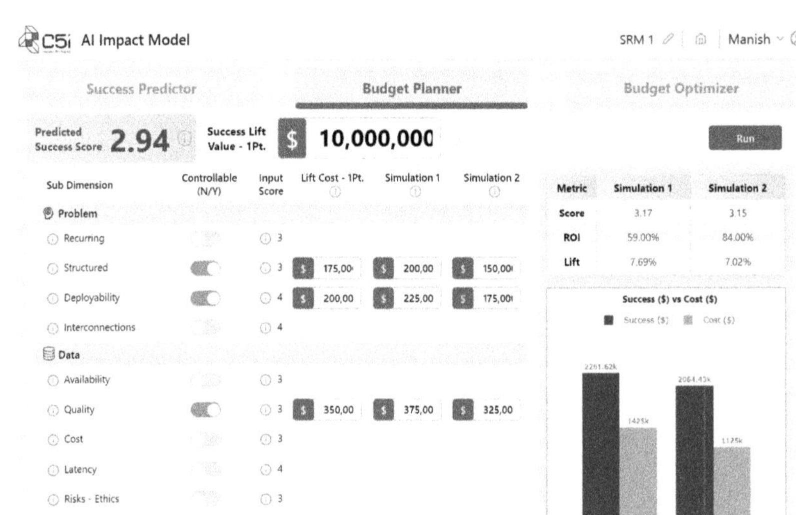

Budget Optimizer

The Budget Optimizer takes decision-making a step further by leveraging AI to recommend optimal allocation strategies tailored to specific objectives, such as maximizing ROI or increasing project success rates. Unlike the Budget Planner, which relies on user-driven inputs and human intelligence, the optimizer uses advanced algorithms to identify focus areas that yield the highest impact with the most efficient use of resources.

Exhibit 9: Changing Budgets to Optimize Business Impact of a Project

Use of the tool has three primary elements:

Data Collection

Firstly, before integrating it, the tool can be used based on the set of learning it has at the back end from programs it's already been used for. The tool can be used as it is today; ideally, however, an organization should bring in some of its own body of work.

They can use some of their own projects and measure them on these eighteen variables, and input that into the tool so it can learn from their own contextual data. The tool has the capacity to learn from its own set of data and from the additional context provided by the organization's projects.

As mentioned above, the AIIM already has data from different projects between various organizations that it has used to create a model. Once organizations planning to use the tool bring in their own data and add it to this model, that will augment the model's data-set. Doing so will make it more relevant and consistent with their business environment. Organizations must understand: how is that data going to get collected? What does the process of data collection entail? Who are the people you need to collect the data from? Send a questionnaire to them and ask them to provide relevant information. Understanding the optimal process of collecting that data in the right manner is step one in using the simulator. Making it more contextual in this way will deliver better results from the tool.

Success Prediction

Once the requisite data has been collected, there are two separate parts to using the simulation tool. The first is success prediction, and the second, covered in the following section, is the optimization functionality. It's important to be very clear with respect to who is going to use the tool. Doing so requires some level of training because you need to understand the eighteen variables and how to rate each variable; you also need to understand the right business questions to ask. So, there should be a certain amount of orientation and training prior to using the simulator. For those reading this book who plan to use the simulator, you're already off to a great start!

It's important to understand how to use the tool effectively and screen out any biases. As described earlier, the success predictor is one element of the tool's simulation engine. This is typically used when you have a new AI program you're considering, and you're able to input the eighteen variables and get a prediction of the level of success you're likely to have. This typically involves a certain business case for which you have established your desired level of impact on whichever metric is important to you (whether it is sales, cost savings, customer experience or something else). You set up an expectation, and basically, if you achieve a 4 on the business impact rating, that means you will likely achieve your business case. If you come in higher, then you will likely do better. If you come in lower, then you're probably going to do worse.

That will give you a prediction of success. Let's say you have twenty programs, and you want to see which ones are going to be more or less successful–the simulator allows you to do that. You can calibrate that with your business case and see what you're going to achieve from that specific project. This is, of course, directional. It's not a perfect science, but it gives you a good sense of direction.

Budget Planning and Optimization

The tool can also be used to optimize existing initiatives. Beyond predicting success, it enables you to actively improve performance. You can assess multiple projects based on their predicted outcomes, identify those you want to strengthen, and then optimize across the eighteen variables to determine where targeted investments will create the greatest impact.

You retain full control over which variables to influence, recognizing that some may be constrained by context or external factors, while others can be actively shaped depending on the situation.

Factors that you naturally cannot change may include certain things about the nature of the project or whether the program is recurring versus non-recurring. It's very unlikely you can change that because that's just the nature of the project that you've undertaken. But there could be other things that you can or cannot change depending on the situation. Maybe your organization has restrictions on purchasing new technology, and they have a certain technology stack that they want you to use; perhaps it's not the best, so you may be stuck with subpar orchestration technology.

In some organizations, you may be able to change the technology being used. Sometimes, you can invest in and improve your data systems, such as by buying new cloud technology or investing in having vendors perform additional data cleansing or other services for you. You might be able to hire some more technical experts, or there may be some people in your organization you can train and turn into citizen AI scientists. On the other hand, maybe people are just not of that mindset, and they don't want to become citizen AI scientists.

Depending on the situation, there may be some variables among the eighteen that you can impact, and some you cannot. The model gives you a chance to say, "Amongst these eighteen, these are the ones that I can improve for this program. This is my cost for improving per point for that variable, this is my total budget, and this is how much I'm planning to spend on these variables." Then, the model will update the success rating and give you a new success score.

Now, we come to the Budget Optimizer.

This module takes things a step further, where the tool's intelligence can be used to dynamically allocate budget across different variables to show the optimal outcome.

In this module, the user would once again specify the variables that they can impact and the cost for improving per point for that variable and specify their budget. The tool will go ahead and dynamically allocate their budget across the ideal mix of variables and showcase the business impact improvement;

this allows the user to identify the best possible investment plan and the outcome or ROI of that plan in terms of business impact from the AI project.

Tips for Using the AI Impact Model Simulator

While we were writing this book, we hosted a client conference and had some great conversations about how teams are using the AIIM. One story in particular stood out. A client described how they brought everyone together for an off-site and asked employees to submit ideas for AI projects. The response was fantastic–so fantastic, in fact, that they ended up with a list of 150 potential projects. As the client put it, "The good news is that everyone was engaged and excited. The problem is, we can only take on about twenty of them. So how do we decide which ones to move forward with?"

What we would do in this case is use the model to, essentially, weight or calibrate each initiative on these variables. It will stack around it and say, "These are the losers to get rid of." That's one application where you can integrate this model into decision-making to find the unproductive or unprofitable programs and kill them. It can tell you which initiatives and projects you should not be pursuing because they're not likely to be successful. That is the first type of application, which rejects projects that are not likely to produce much in the way of return.

When you have a smaller list of, "Here are the twenty projects that I do want to pursue," you can use the simulator to answer the question, "How do I make those more successful?" That's the second application, which is, "What are the levers to push in order to optimize?" So, there's a rejection phase, and then there is an optimization phase for the ones that you have prioritized and have selected to move forward with based on ROI projections.

We use ROI measurements to knock out the dogs. Then, we want to improve the ROI of each project that we pick. That is where you run the simulation to determine, "If I did this, if I did that, if I invested more here or there, what would happen?" The model will tell you, for example, that your success rating is going to go from 3.2 to 3.4, and this is the economic outcome.

The power of the model is that we've calibrated it with over sixty projects. So, you can literally input what your project characteristics look like, and it's going to tell you this is what your likely success is going to be; more importantly, it can help you answer a question like the following: "I thank you for telling me that my project is going to be only moderately successful. What I want to do is to improve the likelihood of the level of success. How do I do that?" The simulator will also allow you to see that, for example, if you invested more in change management and the feedback loop, this is how much lift you would see. Or, if you invested more in data quality, this is the lift you would see. The simulator provides you with the ability to run scenarios and to do *what-if* analysis. This enables you to have a very rich conversation with the model, which sensitizes you to discovering: what are the things that matter most?

Then, the question becomes "How do I make tradeoffs?" "Do I make a tradeoff between data quality or team skills, a dollar here or a dollar there?" Those are the sorts of questions you can answer, and because we have calibrated this model on real data, the recommendations that it produces, the sensitivity analysis and the scenario analysis that you can do are grounded in what actually happens in the real world. The simulator saves organizations the trouble of trying to design their own model to develop project success forecasts. Instead, the simulator provides them with the variables. They just need to tell it what the various rating levels are for the eighteen variables.

Typically, we'll have a team of people from a company providing us with that information in order to calibrate the model. They need to agree that, for example, on data quality for this project we are on a level 3 and not a level 4. That comes from the organization using the simulator. We give them a very detailed explanation, so they know: what does level 3 look like? What does level 4 really mean in practice? Thus, we're not randomly labeling something as a 3 or a 4. We're grounding it in very precise descriptions. "If this is the way you are thinking about this, it's a 3, or if this is the way, it's a 2." Examples are provided to improve the accuracy of the ratings.

Once we've got that data from a group of people across the organization, then we're able to say this was a success. And the beauty of it is that, now, they can say, "OK, I don't like this. I want to do better. I have a million dollars of additional budget. Should I add more people? Should I invest in data quality?" The simulator allows you to make those tradeoffs.

CHAPTER TEN

Conclusions and the Evolution of the AI Impact Model

"One of the most significant impacts that I've seen with the use of AIIM was related to strategic revenue management or revenue growth management. Some of these use cases centered on e-commerce growing significantly as consumer preferences change rapidly, especially post-pandemic. That change in consumer sentiment or consumer behavior has to be captured very quickly, and depending on that, there are opportunities to more effectively promote the product and drive distribution and pricing in the right direction. When you are using AI to accomplish this, you can use the AI Impact Model to compare and contrast different initiatives of this type."

Deepak Jose – Global Data, Analytics and AI Executive. Vice President, Head of Data and Decision Intelligence, Niagara Bottling; formerly served in executive roles at Coca-Cola and Mars

This chapter ties everything together, summarizing the key lessons learned from the AIIM framework and offering actionable takeaways for readers. It emphasizes the interconnected nature of the five Value Levers and their collective role in driving AI project success, thereby paving the way for converting AI's promise to profit.

In the age of intelligent business, adaptation is crucial, especially in the technology sector. With this in mind, the chapter also looks to the future, offering a blueprint for applying the model across industries. This will involve scaling the framework to build industry-specific insights by tailoring it to address the nuances of sectors like healthcare, retail, and manufacturing.

Along with scope expansion to enable the model to address industry-focused contexts, it will also be enhanced through larger datasets and feedback loops. To accomplish this, the model's robustness will be bolstered by incorporating more project data, while establishing feedback loops will help refine its predictive accuracy.

The AI Impact Model is a strategic compass that can help business leaders navigate the complexities of adopting and scaling AI initiatives. In an environment where investments in AI are difficult to measure and their impact even more difficult to predict, the AIIM provides business leaders with the information they need to make informed trade-offs, identify high-impact projects and decide on investment strategies to enhance project success. Additionally, the AIIM provides actionable guidance on balancing trade-offs, such as investing in recurring programs over one-off projects, prioritizing data quality over technology, and fostering in-house expertise.

By applying the AI Impact Model, leaders can optimize resource allocation, enhance team collaboration, and foster a culture of data-driven decision-making. As the case studies presented in this book demonstrate, the AI Impact Model has delivered value in real-world applications, guiding enterprises to deploy AI solutions that are both impactful and scalable.

Before we move on to the Lessons Learned section of this chapter, a quick check-in with our intrepid CIO from the first chapter seems in order. What happened on his third attempt to successfully generate an acceptable return on AI? Our hope would be that, learning from past failures, on the third attempt, he developed a plan that adhered to the points we have made throughout this book.

In taking this scientific approach to creating business impact, he could have benefited from consulting with AI technology experts from both inside and outside the company to draw up and implement a plan. To maximize its chances of success, the plan should take into account the five Value Levers we have covered in the book. Using the AI Impact Model could help him select which of the proposed AI projects were most likely to succeed. Then, he could further optimize the impact of the selected projects by using the AIIM Simulation tool kit.

Armed with this evidence-based approach, the CIO would have been in a good position to generate positive business impact from AI and win plaudits from his CEO. With the company's competitive position bolstered by AI, the Board of Directors would likely be pleased enough to give the CEO some breathing room, which in turn would make her more than happy to support the CIO by providing him with the space and budget to continue driving the company's AI agenda forward.

Lessons Learned

Throughout the book, we've covered the most valuable and actionable lessons we've learned in the process of developing and employing the AIIM. Please note, however, that these are heuristics, or rules of thumb. They are guidelines; thus, exceptions may occur.

In the following section, we present summaries of these lessons so readers can find them all in one place if they need to refer to them quickly.

Structure the Business Problem Well

A well-defined problem serves as the foundation for a successful solution. Identifying the business problem your AI project is designed to fix and specifying how it will work is crucial to success. It's vital to invest sufficient time into making sure that you have accurately articulated the objective, constraints, scope and business value of the problem you are solving for before launching a project. In the case of exceptions to this planning process, like exploratory analyses or proofs of concept (POCs), make sure to approach them with flexibility while still focusing on the value they can create.

Prioritize Projects Where You Can Deploy at Scale

The return on AI initiatives often hinges on whether they can move beyond isolated pilots and into scalable deployment. AI projects that succeed in seamlessly connecting systems of insight (analytics and intelligence) with systems of records and systems of engagement/ action improve their ability to drive meaningful actions across various user personas. This approach boosts the chances that an AI project will be deployable throughout an enterprise's IT ecosystem.

Data Quality Is Critical, While Data Availability Is Important but Has Diminishing Returns

As stated previously, poor data quality can torpedo an AI project all by itself, so validating that the data you will be using is high-quality should be an essential part of the planning process. Avoiding the "garbage in garbage out" phenomenon is crucial to implementing effective projects.

While data availability is important, diminishing returns apply after a certain point. Fixating on gathering every piece of data possible before starting can result in delays that erode business impact and ROI. Additionally, larger and more diverse data-sets can increase latency and cost. A more pragmatic approach would be to begin with a minimally viable dataset and refine models iteratively as more data becomes available. It's important to keep in mind that the data-set being used is representative of that stream of data. This enables programs to start delivering value sooner while maintaining flexibility for future additions.

Privacy Risks and Ethics are Table-Stakes Issues

While failing to safeguard customers' private information or failing to adhere to ethical policies in collecting data can sink a project fast, taking these steps will not necessarily make a project successful. Privacy, risk management, and ethical considerations are necessary but not sufficient to guarantee the success of AI projects. These foundational "table stakes" issues must be addressed as prerequisites, but they do not, on their own, ensure positive outcomes. Organizations must safeguard these factors yet look beyond them to create value from AI programs.

Culture, Change Management, and Feedback Loop Have Important but Indirect Effects

Certain variables might be less significant by themselves but collectively exert a powerful influence on the success or failure of data, analytics, and AI programs. An organization's AI adoption culture or change management policies can play a major part in generating results. These variables by themselves typically don't make a project successful. Instead, their interaction with other factors associated with an AI project, such as data quality, technology, leadership support, etc., can indirectly help a project achieve its objectives. AI initiatives benefit from robust change management policies such as clear communication, alignment across teams, and proactive strategies to address resistance to change.

It's also important to build a feedback loop that collects feedback from users and model performance data and feeds it back to enhance the original model. Another key element in many successful AI initiatives is a strong AI adoption culture. This includes embedding data-driven decision-making into the daily operations of the benefactor business and ensuring that insights generated by AI are actively used to guide processes and strategies. Teams must be trained on effectively partnering with GenAI and agentic AI tools to augment their work. Additionally, fears related to job replacement should be addressed up front. By focusing on these interconnected elements, organizations can significantly enhance their ability to deliver impactful, scalable, and sustainable AI programs.

"From my perspective, the two most important takeaways from using the AI Impact Model were, first, change management: don't just rely on technology and lead with technology, but really think about change management to make a project successful. Focus on adoption because that's incredibly important to successfully launching a project. Second, make sure, as you're thinking about some of these measurement techniques or thinking about these projects, to have a data scientist or somebody who really understands the model to be part of the process so that you can set the parameters up correctly to measure impact."

Ajit Sivadasan – Ex VP and GM,
Global eCommerce, Lenovo

Prioritize Recurring Problems

While it can be tempting to look for AI projects that grab people's attention, the most productive AI projects from the standpoint of the business case often involve solving recurring problems. Focusing on recurring programs–those with predictable patterns and repeatable processes–generates sustained value and higher success rates. While one-off or exploratory projects can offer learning opportunities, recurring programs are better suited for automation and scale, driving consistent ROI. The AI Impact Model emphasizes this by highlighting the importance of aligning AI investments with business processes that deliver ongoing impact rather than transient gains.

Focus on Data over Technology

Our research shows that high-quality data is crucial to AI project success, while technology is important but not necessarily determinative. As a result, if given a choice between the two, typically, investing in high quality data for a project should take priority over investing in technology. Data often has a greater impact on success than the technology stack itself. The AI Impact Model encourages prioritizing investments in robust data pipelines and governance before selecting or building AI technologies. Technology can be upgraded or replaced, but without accurate, reliable, and well-structured data, even the most promising AI projects are likely to fail.

Invest in Technical Expertise over Resource Availability

While it's usually useful to have access to consulting firms and similar resources that can help a company deploy an AI project, this should not supplant investing in a firm's own technical expertise, at least to a limited extent. It typically pays dividends to have some level of in-house expertise that can be used to vet and propose AI projects. With technology changing so rapidly these days, building some level of in-house technical expertise typically offers greater strategic advantages than resource availability. Internal teams possess a deeper understanding of the business context and can iterate on solutions more effectively.

The AI Impact Model underscores that investing in some technical talent at the client's end enhances the quality of AI implementations. However, the model also highlights that having substantial in-house resources is not a driver of success. This demonstrates that the ideal approach for most organizations is to have some technical capabilities in-house and then partner with vendors who can supplement resources and bring their complementary expertise and knowledge to a project, leading to the ideal combination.

Aside from Certain Scenarios, Cost of Data Typically Matters More than Latency

While low-latency data processing can be critical in some scenarios, such as real-time decision-making, the broader cost implications of data acquisition, storage, and processing often have a more significant impact on the overall ROI of AI initiatives. The AI Impact Model helps businesses weigh the relative importance of these factors, guiding them to prioritize data strategies that align with cost-efficiency and long-term sustainability while only optimizing for latency where it delivers substantial value.

By applying the AI Impact Model to these trade-offs, organizations can make informed, strategic decisions that maximize the impact of their AI and analytics programs while navigating the constraints and complexities of real-world implementation.

Further Development of the AI Impact Model

By pursuing a variety of development directions, we are positioning the AI Impact Model to become an even more powerful tool for business leaders, equipping them to harness the transformative potential of AI while staying ahead of the curve in an increasingly competitive and dynamic landscape.

In the remainder of the chapter, we explore the various ways in which the AI Impact model is being enhanced to expand its capabilities and applicability. There are three primary avenues we are pursuing in this regard:

> **Incorporating more Generative AI Projects**: At the time we conducted the empirical research, generative AI programs were still in the stage of planning and deployment. As time has gone by, we have increasingly included data from concluded generative AI along with traditional AI/ML projects.

Diversifying the Dataset: To enhance the model's accuracy and relevance, we have extended our research to include a wider range of projects and clients. Partnering with a broad spectrum of clients will provide greater accuracy in the applicability of the model to different sizes and types of companies.

Addressing Industry-Specific Nuances: The AI Impact Model can be adapted by incorporating industry-specific projects to generate more specific insights for business leaders in industries like healthcare, finance, or retail. While the model itself is valid across industries, the relative importance of the Value Levers is likely to be different across industries.

The following sections offer a detailed exploration of how we are pursuing measures to enhance the model's functionality in response to client requests and the further development of AI functionality.

Expanding the Data-set and Using GenAI for Descriptions

We are now expanding the data-set underlying the AIIM by bringing in more programs; at the same time, it is being implemented by a number of our clients. Over time, we plan to create industry-specific versions, where the model will learn from the overall data set, and will also learn from data specifically relating to an industry. We may also explore creating function-specific sets, for instance, for marketing, for supply chain or for other factors such as organizational risk, etc.

Anyone who follows the sector knows, of course, that much is changing in the world of AI. This change has been led in the past few years by generative AI, now joined by agentic AI, and that will continue to influence the evolution of the model. We plan to ensure that it advances along with these developments in order to work with all types of AI.

For instance, we are currently evaluating a way to receive a project description from somebody for our AI Impact Model and help them populate the necessary fields. We are creating a generative AI system that can actually read that project description and use it to populate some of the ratings of the model. So, it's generative AI for AI. First, we were using AI for AI, and now we're exploring using generative AI for AI. In this application, we're creating an agentic GenAI system for the purpose of reading a project description to help provide information on variables to the model.

AI Type and Factor Performance

Currently, the three general types of AI projects are traditional machine learning, generative AI, and agentic AI, the latest entrant to the field. One question that people often ask us is, "Do you think that when we look at generative AI projects, we're going to find that there's something different about them, that there are different factors that will be responsible for success or failure?" Our take on the issue is that we think the variables and the factors–with the factors being the five Value Levers, and the variables being the eighteen subdimensions–that make up our framework are universal. They apply across different types of AI.

What may be different is, for example, the relative importance of problem scoping or framing for generative AI. The data lever may have some changes in weighting due to the use of large-scale unstructured data. We will have more to say about how the latest developments in AI may affect the AIIM in Chapter Eleven.

Developing the AI Impact Model Across Industries and AI Types

"I think at some level we should be leveraging the AI impact model for teaching students from high school to college to the grad school level. Students who are focusing on actually becoming AI practitioners because there are a lot of books that will talk about why this or that technology is better. Why this mindset is better or this framework is better, but nobody talks about change management connected to AI in a comprehensive way. So, I do see the value of leveraging the AI impact model across the board."

Deepak Jose – Global Data, Analytics and AI Executive. Vice President, Head of Data and Decision Intelligence, Niagara Bottling; formerly served in executive roles at Coca-Cola and Mars

We are in the midst of further research and development of the model in two directions: one is across industries, and the second direction is across types of AI. The goal is to enrich the model and allow it to make more prescriptive statements that are industry-specific or AI-specific–that's where we are heading with the model.

We frequently get questions from people in different industries, such as the following: "I'm in health care or in financial services, are there finer nuances I should keep in mind when applying the AIIM?" The answer is, yes, there will be industry-specific differences. At this point we can't make those statements because we don't have enough data from different industries represented in the data-set. But let's say that we had five hundred data points from ten industries. In that case, we can say something like, "In insurance, this matters more, but in consumer-packaged goods, that matters more." We are working on gathering this data in our current development phase of AIIM. So, we can say where the project is doing that, and that we're now working on getting more data to help further refine its ability to answer such questions.

Generative AI is yet to deliver meaningful enterprise value for most organizations, and agentic AI is at an early stage at the time of the writing of this book. In order to develop the model further and prepare it for an agentic AI project analysis, we will need more data on projects of this type. We are in the process of collecting more data so that we can simulate these projects effectively. The beauty of a framework like the AIIM is that it can be used across different contexts, and we believe this includes generative and agentic AI. Defining the problem and the other Value Levers will very much apply to these kinds of projects.

For agentic AI, it could be that we will extend the model to deal with new ideas such as agentic orchestration, where you have AI agents controlling other AI agents to manage full workflows. In the final chapter of the book, we will further explore this idea of agentic orchestration and business process reengineering. This includes evaluating the possibility that there may be some new functionality around that, which will involve bringing in new flavors with agentic approaches as well as taking a deeper look at what the rise of generative AI may mean for the model. Interestingly, on some initial work we have done, our model is performing reasonably well at predicting success in early-stage agentic AI projects.

CHAPTER ELEVEN

The Future of AIIM with Generative AI and Agentic AI

The excitement in recent years over first generative AI and now agentic AI stems from the realization that these technologies represent the next generation of AI development. In this chapter, we'll take a broad look at how they will influence the future of AI-led business impact and how this might impact the AIIM. This chapter takes a conceptual and even, at times, speculative approach in order to give readers a glimpse into the authors' thoughts on these topics.

Generative AI has more of an impact on a broader set of businesspeople than traditional machine learning AI. The reason for this is that GenAI touches many areas not previously covered by traditional AI applications, such as creating marketing imagery, drafting communications, and even advising on strategy. Machine learning applications are typically focused more on quantitative data related to sales, marketing, operations, supply chains, finance, etc. While agentic AI is a more recent development, its potential for highly autonomous AI functionality is starting to capture more and more interest both from organizations and industry experts.

Overall, this chapter serves as both a conclusion to the book and a forward-looking vision for how the AIIM framework will continue to evolve as an indispensable tool for empowering organizations to unlock the transformative potential of AI.

The AIIM and Advances in AI

As of the current moment, from the initial findings we have seen, variables that impact generative AI success do seem to follow similar trends as traditional AI, but that could change in the future. For instance, as mentioned earlier, defining the problem may have less weight because the AI can help you do this. Agentic AI, which we'll discuss later in the chapter, takes things to the next level because now it uses generative AI along with access to the company's machine learning models, tools and platforms to complete actions. We can now, quite seamlessly, link Data to Insights to Action. In Chapter Ten, we discussed how the AIIM will need to adapt to include these types of AI in its analysis. This extensibility is supported by the framework underlying the AIIM, which allows it to be transported across contexts and applied to different problems.

Essentially, the principles of getting the problem accurately defined, picking the right tools, leading the change manage-ment, etc., don't change. Therefore, we are quite confident that the fundamental machinery that we have is sound and will be applicable to generative and agentic AI. However, the individual weights of the different variables and factors may be different.

Generative AI

Generative AI is becoming a more mature technology. By the time this book is published, it will have been publicly available for three years or more. There is now a growing number of generative AI implementations. We have already included quite a few in the AIIM model, and this will continue to grow, giving it a larger role and greater influence on the model in the future.

If problem definition becomes easier with generative AI because you can use AI for this purpose, paradoxically, maybe problem definition becomes less important as a variable because now everybody has a higher chance of getting it right.

On the other hand, there may be some new data governance challenges, because now we're dealing with unstructured data and other complicated types of data. So, data issues may become more salient. The relative importance of the factors and variables is in driving the success of the project will perhaps be different, and we don't know exactly how that will develop, but it doesn't change the underlying structure or validity of the model itself.

In the following sections, we will explore these and other issues relating to the impact generative AI may have on the functioning of some of the factors and variables in the AI Impact Model.

Data

One major development brought about by generative AI has to do with data. Data quality and data availability are still going to be important, but if you have a certain amount of data available, you will now be able to use that to generate synthetic data. One of the things that is very interesting about some of these foundational AI systems is that they can actually generate very effective synthetic data. But you can't generate synthetic data by itself. You can only generate high-quality synthetic data if you already have a good base of existing data.

This goes back to the point that you can solve for data availability if you don't have data available beyond a certain point. If you have a certain amount of high-quality data, then you can generate synthetic data to augment it.

Systems of Insights and Orchestration

Another development centers on systems of insights, or insights platforms. The expectation from insights platforms and how insights are consumed is completely changing with GenAI. Now, the expectation is that your insights platform is not just a dashboard but is available to people on the go;

they should be able to talk to it in a natural manner and get back intelligent, contextual responses. That involves the use of generative and certain agentic systems to understand your question, go and query the data, and come back to you with the answer after running the analytics. The same thing with regard to deployability will also be relevant for orchestration technology. AI agents will be able to stitch seamlessly from your system of insights to your system of action and also create a feedback loop.

Defining the Problem

If we go through the different causal factors in the context of problem definition, with traditional machine learning, the problem is actually fairly precise and quantitative. You mostly know what model you want to build, and while it is important to define the model correctly and to define the problem correctly, it's still a structured problem. But in the case of generative AI, the answers are only as good as the questions you ask. It reminds one of the quote by Voltaire that, to paraphrase, *a person should be judged by their questions, not their answers.* If we modernize the quote, we might say, *judge a prompt engineer by their prompts, not the answers.*

In problem definition, now the question really becomes an issue of prompt design; it centers on what questions we're going to ask the model. In prompt engineering, there are now best practices surrounding how you structure prompts. There's a framework that we call RCTO, which is role, context, tone, and output. You're defining all these guardrails, and what we're not talking about enough beyond prompt engineering is context engineering.

Context engineering is the idea that you not only provide the model with a prompt that is intelligent, but you also provide it with all the context that is needed for it to provide you with the right answer. For example, providing the model with proprietary corporate data, providing the model with access to tools, etc.; essentially, setting the model up for success. That is context engineering, which is a broader notion than prompt engineering.

All of this adds up to the fact that defining the problem takes on the mantle of defining the questions, defining the context, defining the prompts, and figuring out the ways in which that is robust. The interesting part here is that if you don't have a well-defined problem, you can use good context engineering to define the problem more effectively in terms of objectives, constraints, success outcomes, etc. In such cases, problem definition takes on a different shape and may become more about context engineering. So, its importance may remain as high, but the "how" may evolve.

This takes us back to the observation that the variables themselves won't change in the context of generative AI, but the weights may change. In this light, perhaps problem definition (or context engineering) actually becomes more important rather than less. Due to the unstructured nature of the problem, there's a lot of skill involved in prompting and in defining.

Talent

In an earlier chapter, we emphasized the importance of having technical talent. But with generative AI, less skill is needed to interact with AI models because you can now ask them natural language questions. Thus, generative AI is a democratizing force. Therefore, to the extent that is true, perhaps the importance of having highly skilled AI talent is less than for traditional machine learning projects. That's one possibility.

Generative AI is leading us to a no-code, low-code revolution. While there will be a need for some highly technical people to build the AI and deploy the AI models and associated technologies, the much greater need will be to understand "enough" about AI and the associated business dynamics. Therefore, the need for citizen AI scientists will continue to grow. To quote Ashwin Mittal, one of the authors of this book: "The technology has evolved so much and so fast that the gap between the AI experts and novices is not as wide as it was before. If you want to be a citizen AI scientist, now you don't have to have had ten years of experience dealing with AI. You just need to be smart, curious, and adaptable."

Risks: Governance

With generative AI, it's important to build guardrails and enact appropriate governance to make sure that the AI is painting within the lines, because the problem with generative AI is that the output it generates is inherently probabilistic. It's not deterministic. You're never a hundred percent sure that it's right. So, it's important to create the right guardrails, gover-nance, and human oversight over generative AI models. This is different in the case of traditional AI and machine learning. With traditional AI and machine learning, outputs can still be probabilistic, but that is most well-known–e.g., you are predict-ing the likelihood of a customer buying a product at a specific price or predicting the number of units that will be sold of a specific SKU next month. With generative AI, its probabilistic nature leads to the generation of content, which the user sees not as a prediction but as a usable output.

Also, since generative AI can be used for certain cognitive or strategic decisions, like medical diagnosis or deciding whether to engage in litigation and so on, we have to make sure that it's not wrong. The risks of being wrong are greater now than maybe when you're monitoring a turbine's behavior using traditional machine learning and using that to do predictive maintenance. We think that there are some higher-risk scenarios where the consequences of a wrong decision may be greater and, therefore, the importance of putting the right guardrails and governance in place and monitoring that on an ongoing basis becomes even greater.

AI gets more powerful with generative AI, so the stakes get higher. AI governance includes governance designed to keep the data secure, keep it private, and so on, but now it also focuses on how to reduce/avoid hallucinations and ensure that models don't spout offensive or abusive remarks. So, AI governance includes data governance, but it's a broader idea. In our AIIM, the sub-lever around risks and ethics is in our Data lever. Going forward, we will have to view risks more broadly beyond data to overall AI risks.

Agentic AI and Autonomous Systems

When we get to agentic AI, it is possible, as mentioned previously, that we may need an extension of the model. The biggest change here is agentic orchestration. The idea behind agentic orchestration is that you will have agents working with other agents, multi-agents, and these agents will need to be supervised by supervisory agents; then, you end up building this whole agentic organization. We first discussed this in the chapter covering systems of orchestration.

In agentic AI, the major leap being made is that the AI now starts to not only make predictions, but it also makes decisions and takes actions. It starts to actually perform workflows and tasks in an autonomous or semi-autonomous manner. It should be noted that agentic AI is not always autonomous; there might be a human in the loop. This could occur in the last mile or at specified points along the way. Generally, agentic AI functions by seamlessly linking together your systems of record, insights, and engagement. The question is, how does the world look different with agentic AI? Given the relative novelty of the technology, the authors' thoughts on this matter must necessarily be considered at least somewhat speculative.

One view of the future development of agentic AI is that we will ultimately see, for many processes, agents orchestrating instead of people–agentic orchestration. There is also something called agentic mesh, which is basically a mesh of agents working on various tasks. In the future, we see the evolution of what can be called agentic ecosystems. An agentic ecosystem is the idea that, first of all, you have individual tasks being performed by agents. Then, as you start thinking about workflows, the agents may have to coordinate with other agents. The next level is to chain together these tasks and build agents that can supervise workflows and entire business processes. By the way, in all of these cases, the other point to make is that the agents will not necessarily work autonomously. They'll work with appropriate human intervention and human oversight. Thus, it's not like everything is running on autopilot.

For example, if you have an agent monitoring what's happening with the competition and it detects some unusual competitive activity, that agent may go talk to the customer agent and say, "Hey, is this a relevant trend for customers?" Then, that agent may talk to the product road mapping agent, saying, "If that is important, let's put this on the product roadmap." It's almost like we have co-workers talking to each other. The agents will talk to each other, and they will talk to human beings.

The next level of agentic AI is the agentic ecosystem, where our agents now transcend the four walls of our enterprise, so we can, for instance, imagine a scenario where you are an automotive manufacturer, and you want to buy 10,000 tons of automotive-grade steel. To accomplish this, you might send out a seller agent who is endowed with all the parameters. "This is how much we want to pay; these are our quality terms; these are our financing terms." It has all of the information it needs for the negotiation.

That buyer agent then goes out and says, "Hello, are there any seller agents who are bidding for this business?" The seller agents and buyer agents will meet in the ether. They'll meet in some sort of agentic space. Now, when a seller agent talks to a buyer agent, they have to authenticate and say, "First, prove to me that you are an authentic seller agent, and by the way, I'm also going to do some due diligence on you. Prove to me that you have the capacity, prove to me that you will protect my data, prove to me you have the financial wherewithal." Once those qualification criteria are met, they can lock in and say, "OK, now we're connected, buyer and seller agent." You might do this with multiple seller agents, and then the buyer agent may conduct a reverse auction, saying, "OK, bid, this is 10,000 tons of steel I want. There are three qualified suppliers. You tell me the price." And it may conduct that auction and then place the order with the supplier that best meets the criteria.

This is a futuristic scenario. It could happen in the next couple of years, but this will, in turn, require protocols and standards. Agents need to be able to talk to agents, authenticate, and make sure that they are exchanging information in a secure manner. This is what Google's early version of A to A (agent-to-agent) protocol is about. Until those things happen, you won't be able to have agents go across your company to other companies. Ultimately, an agentic ecosystem means that your agents will be talking to your customers' agents, your suppliers' agents, and your partners' agents. There will be a whole new class of commerce that we call agentic commerce; we've talked about B2B commerce. This is A2A–agent-to-agent–commerce.

How will the model handle this? In that context, there may be some new phenomena like agentic security and agentic definition, since, from a problem definition standpoint, we might have a n agentic workflow definition. There might be extensions, and similarly, from a tooling standpoint and data standpoint, there may be new challenges and new opportunities. We think agentic AI may push the boundaries of our model and invite extensions that will accommodate some of the new class of phenomena that will be introduced with this agentic governance.

We believe that in the context of agentic AI, we will likely need to build further functionality in AIIM. Why can't we do that now? The reason is that we don't even know exactly how these agentic workflows will be actualized.

As a result, we need a bit more time for the dust to settle, but we believe that this is going to be a much bigger and more impactful development than the impact thus far of generative AI by itself.

It might be said that we will need a new HR organization for digital labor orchestration. Basically, similarly to human employees, agents will need to be recruited. They will need to be trained and onboarded, and just like employees, agents will need to be monitored. Additionally, agents will need to be paid such that every time the agent is running, there are inference costs and computation costs. That's the salary of the agent. If we apply the entire HR end-to-end framework for agentic, we might say that we have a vision that in the future, the head of HR and the head of IT may be the same person! Because now you're managing a blended workforce, consisting of human employees and agents, and you will also need to look at roles and paths and divide them.

Between agents and humans working together, it's almost like you have digital co-workers. All this gets at a potential new paradigm in human resource management, which is not digital human resource management, and in fact, we may not even talk about head count. We might talk about blended count or head count plus agent count, and one of the metrics that we might think about is the percentage of our work that is done by agents. That's like an agent utilization percentage, and another thing that might be measured could be a human augmentation factor, which is the leverage one is getting on human employees through the use of agents. What productivity is the average human employee achieving with the use of agents in your organization compared to what they were achieving without agents? What is that leverage I'm getting? That may distinguish companies. For instance, "I'm at 1.4, you're at 1.7." Basically, the agents will augment and amplify human capacity, and we will become more productive. So, there might be new metrics we'll need to create to measure the impact of agents on productivity, on output and outcomes.

In the following sections, we take a look at the impact agentic AI may have on the functioning of some of the factors in the AI Impact Model.

Deployability

If we think about agentic AI and what it can offer, one of the most potentially powerful functions it can serve is to stitch things together very effectively from systems of records to systems of insights to systems of action and engagement. Thus, agentic AI, if used well, can solve to a large extent for deployability. Organizations that are effectively using agentic systems will be able to score much higher on deployability. This is an important development in AI efficacy. Improved deployability will create a gap and a competitive advantage in ROI from AI programs for such organizations. So, the importance of deployability will not decline, but at the same time it will become much easier for some organizations to score highly in this category.

Problem Definition

Problem definition takes on a different meaning with agentic and autonomous AI because it is now defining the workflows that you want to automate. It's defining the jobs to be done, the steps the workforce must take.

For example, one of the authors was speaking to a group of surgeons about clinical operations, including the patient journey through surgery, right up to the end process. This involves patient access and intake. Then, there is the preoperative planning. After that is the preparation. Then, there is the intra-surgery support. Next, there are pathology labs and tests that happen after the surgery. Then, there is the postoperative recovery and rehabilitation, and after that, there is the patient discharge. Those are the steps in the process. With agentic AI, we're going to first have to understand the sequence of tasks and jobs to be done in a business, in a workflow, as in the example just given of clinical operations.

Thus, we first need to map the workflows, map the sequence of operations, map the tasks, map the activities, and the jobs to be done, then look at how AI can perform each of those tasks; then, also ask in each of these subtasks or tasks, can AI operate in one of three modes. These are: AI either assists the human, augments the human, or automates the human. We call this AAA. In one case, the human is in charge, and AI assists. In the second case, it is AI augmenting the human, and then there is automation, in which AI leads, and the human just provides oversight.

The type of AI you use will depend on the use case. For example, when you're performing the surgery, the human has to lead. But if you're doing paperwork in preparation for discharging the patient, as this is an administrative process, the AI can take the lead. So, in the context of agentic AI, problem definition means defining and structuring the workflow, jobs to be done, and the task to be performed; then, for each of the tasks or subtasks, deciding the level of human AI contribution and balance and deciding how to stitch all these tasks together into a larger workflow. Thus, with agentic AI, it seems likely that problem structuring and definition become even more challenging. One of the biggest differentiators for successful agentic AI programs will revolve around problem definition.

Using agentic AI is like stitching 100 models together in a set of decisions. And the stakes are also higher because now you are making decisions and taking actions. You're not just making a prediction. The first progression is from prediction to creation–which is generative AI–and the next is to autonomy, which agentic AI offers.

Talent

Another interesting thing with regard to agentic AI will be what types of teams does it involve? The surgeons in the earlier example were asking who should be on the team to design an AI project in agentic AI. The answer is that somebody on the team has to understand the workflow. Somebody has to understand the domain. So, you need a physician. But then somebody has to understand data science and AI. Somebody has to understand your enterprise IT system because until this thing connects with agentic AI and your core enterprise system for electronic health records, it's not going to work.

You need a combination of IT expertise, AI expertise, process and workflow expertise, and domain expertise. So, the way you structure the team and the people that are involved is likely to become even more challenging in the context of agentic AI. The role of the citizen AI scientist will need to evolve to include core Business Process Reengineering skills.

Risks: Governance

Another thing that will become very challenging in the context of agentic AI is governance. With agentic AI, defining security, safety, the problem, and who is performing which task takes on great importance. The question becomes, who's liable if you make a wrong decision? If the AI performs a task and the task is performed incorrectly, things can get complicated quite quickly. For instance, in a robo-taxi, if somebody gets killed, who's at fault? Who's liable? There will be new questions that need to be asked in the context of governance.

In the context of generative AI, because decisions weren't being made, we didn't need to worry as much about security, compliance, and data privacy; we just wanted to make sure the model didn't hallucinate. But it's one thing for a model to hallucinate and give you a wrong prediction or a wrong text, but something else altogether for an agent to hallucinate and make a wrong decision or directly use content that includes hallucinations in communicating with others. That may have significant consequences, including liability, customer loss, financial downside and reputation issues. This takes forward the idea of broader AI Governance beyond data that we highlighted earlier.

Change Management and Feedback Loop

Our assessment is that change management and the feedback loop will become even more critical with the use of agentic systems. The reason for this is that the key with agentic systems is to figure out what roles your agents will play and what part of the workflow you will introduce agents into. Then, you must determine what the right agent/human combination is. In order to do that, you need very good process engineering. You need to understand how your process works, and then you need to reengineer processes effectively to figure out how they will run post using agents.

You may not have to invest as much in new technology, but you will likely need to invest much more in change management, partly because it's so crucial to get it right. If you're taking the human out of part of the process, you've got to monitor that or have some way to change that without getting numerous errors. Change management becomes more challenging because now you're talking about replacing humans. You're talking about changing the fabric of work and what work is performed, who performs the work and what skills are needed to perform the work. The reason is that now you will have this very fascinating notion of AI co-workers, digital co-workers, agents who are working side by side with human beings. And that can be a very uncomfortable notion–that now an agent might be your co-worker, and you might have agent supervisors, or you might be supervising agents.

AI Adoption Culture

Along with change management & feedback loop, AI adoption culture becomes even more important with agentic AI. To those working with AI agents, it might seem, in some sense, as if a science fiction novel or movie has come to life. It's almost like you have your digital buddy at your side, and now you need to know how to work with them. Thus, adoption of agentic AI will be constrained by these concerns, fears and risks that people perceive because now AI is starting to do the work of humans.

People might say. "This is what I used to do. What do you mean by telling me an agent will do it? This is my job, so what happens to me?" There is likely to be significant job displacement with agentic AI–not necessarily job replacement, but displacement. Humans will be needed to do higher-order work. Humans will become less executors and more orchestrators. They will need to orchestrate workflows. They'll need to do the design. They'll need to do strategy, but they won't be actually executing the work as much–that's where AI will take over.

But what happens if your life has been built around execution, if that's all you know how to do? And this word orchestration might be seen as not only a big and ugly word, but also as something you don't understand, that you're not comfortable with. There is likely to be a lot of pushback from people who cannot change or don't want to change, and don't want to let agents onto their turf. This is where AI adoption culture becomes vital to the success of projects. Success in this area will likely involve changing incentives. Organizations will have to mandate certain initiatives to accommodate this new way of working.

Leadership Support

As discussed above, change management with a feedback loop and AI adoption culture are likely to become far more challenging with agentic AI because now you're getting into the heart of what people actually do at work. Organizations that neglect collaborative engagement with their staff when adopting agentic AI run the risk of people simply not engaging with the technology at best, and at worst, actively impeding it. This speaks to the need for strong leadership support.

Leadership buy-in and initiative in spreading the word about the importance of getting behind a project are vital when adopting agentic AI. Executive sponsorship is likely to become more important in agentic AI because of the need to drive adoption from the top. Leaders have to convince people of the benefits of changing the structure and function of the organization, as well as its business process and workflows, along with changing the role that humans play in relation to AI; that has to be ingrained into the thought process and the culture of the company. Setting that process in motion has to come from the top down.

The Autonomous Enterprise

If we take the idea of AI agent orchestration to its logical terminus, we arrive at what is called the autonomous enterprise. Dr. Sawhney wrote a book on the topic in 2017 called *The Sentient Enterprise*.[9] The book covered the idea of the sentient enterprise based on the technology available at that time. As a thought experiment, it speculated on the creation of the enterprise as a sentient organism, one that could sense, interpret, and respond autonomously at scale with minimal or no human intervention. That was our thinking eight years ago. And that's where the book posited that things were heading ultimately–a world in which we will have agentic organizations, agentic enterprises, and agentic ecosystems where agents will be interacting with agents. This would naturally include agent-to-agent or A2A commerce. These agents will run autonomously, where human beings only provide the parameters guiding their activities.

The timing of the emergence of agentic ecosystems consisting of entities working together autonomously on a large scale is difficult to predict. We've mentioned it in this chapter primarily as a nod to one of the directions in which we see AI heading. Certainly, there is a great deal of uncertainty involved when it comes to predicting exactly how and when we might get to the sentient enterprise.

Nonetheless, given the rapid development we've seen in AI technology in recent years, it's an interesting question to ponder with regard to potential future developments in the sector.

The Future of the AI Impact Model

We've engaged in some conceptual speculation in this chapter; the dust will need to settle before we know exactly how future AI advances will play out. It may be that there are new questions. In fact, we might need to restructure the model so that it can now say: "What if we add on to the AI impact model something called an agentic impact model?" Which is to say, if I'm launching this agentic project, how does this amplify the productivity of my enterprise? We might at some point need an extension to the AI Impact model that looks at the impact of agentification, or agentification potential, and the outcomes that we see in terms of amplification of productivity and improved efficiency.

If that occurs, it's almost like we're building a theory of relativity so that we have a new phenomenon because now we're unifying space and time, so it gets outside the parameters of the boundaries of classical Newtonian physics. Similarly, here, we get beyond the classical way of considering what the model does, and it might be a very interesting future to think about agentic impact and create a model to measure it. But the contours of that future are ill-defined currently–it is so recent that it's hard to wrap arms around what that might look like. That being said, if we were to speculate, we certainly believe there is enough of a kernel of the idea indicating where the future is taking us.

The model, of course, is fully ready for such an eventuality. We will adapt it to deal with however the future goes once it becomes concrete enough that it's feasible to make the appropriate changes. In that regard, it's important to make a distinction as follows: do the variables change, or do the weights of the variables change? We believe the weights of the variables will absolutely change, but do the underlying five factors and eighteen variables change? We think those are expansive and comprehensive enough to accommodate generative and agentic AI. Some of the variables might need to be thought of differently, as we have highlighted with some examples so far.

We've discussed our thoughts on how the weights might change, but this is speculative until more data comes in. We don't know until we actually calibrate, but we could propose a hypothesis–based on the analysis presented earlier in the chapter–that, for example, to the extent that generative AI will impact more workflows and more departments and potentially have a broader impact than traditional machine learning, change management with a feedback loop may become more important. At the same time, structuring the problem may become more important with agentic AI, while GenAI itself provides help in problem definition. Similarly, change management with a feedback loop becomes more significant with agentic AI because now we're talking about change that is broader and change that affects more people.

Beyond making some directional hypotheses about how the variables might change, our position is that the truth about this really is an empirical question. It's only data that can tell us. These are some speculations that we might make in terms of how, for the model, the relative importance of the variables may look and how the interaction effects might change, but the jury is still out on exactly how they will change. We'll have to wait until we actually put the data into the model to find out.

It would not be profitable at this point to speculate and say we're confident that variable x becomes more important than variable y, etc., because we don't quite know yet. What we have done in this chapter is to offer some informed guesses and some directional thoughts on how this might work, such as our analyses of how emerging developments in AI are likely to affect various factors of the model. One further point to make in this regard is that the model is generalizable and can be applied broadly across different use cases.

While the exact direction the future of AI will take is impossible to predict precisely, the old saying still applies: *a journey of 1,000 miles begins with a single step*. It is important today to start evaluating the impact of what you are currently working on and how it might be impacted by the advances we are seeing in AI. It's about mastering the present while positioning yourself for the future. If you do both, you will be poised to benefit from the powerful productivity enhancements AI can deliver currently as well as in the future. Another saying is relevant in this regard: *if you keep your head in the clouds and your feet on the ground, you will become a very tall person*. Look to the future while staying grounded in what is possible in the present.

Taking this approach avoids the trap of either extreme–that we're paying attention to today, but we're not looking at the fact that even this is getting disrupted. On the other hand, you can't just sit and speculate about the future of the singularity of super-intelligent AI because you have work to do today. It's essential to focus on business impact and ROI today. Thus, it's a combination aimed at doing both right: looking at the hard realities and data today, but also making sure that your organization is positioned for what the future might bring.

When thinking about the latter, we can certainly say that in the AI realm, and to some degree, technology as a whole, the issue is that the future seems to keep accelerating. To cite another quote from Dr. Sawhney, "We're finding the future isn't where it used to be, it's gotten closer." With technological change accelerating, it pays to stay on your toes and be ready to adjust your strategy rapidly if needed. For instance, since we started this modeling work, generative AI has become widespread; then, just as we were getting our arms around generative AI and its possibilities and the use cases and the governance challenges, agentic AI suddenly appeared.

And who can say what's next? Perhaps AI powered by quantum computing? The takeaway is that as organizations and as business leaders, we have to increase our clock speed and improve our ability to absorb new information and quickly deploy solutions. This is why we think the AIIM is an extremely useful tool for navigating the challenge of generating return on AI, now and in the future.

TURN AI INSIGHT INTO MEASURABLE IMPACT

Step into an interactive environment where you can test scenarios, input your own data, and see how AI initiatives can be evaluated, prioritized, and optimized for business results.

EXPLORE THE AI IMPACT MODEL

https://www.c5i.ai/ai-impact-simulator/

ABOUT THE AUTHORS

Ashwin Mittal

Ashwin Mittal is the Chairman and Co-Founder of C5i, a renowned analytics and Applied AI consulting firm that has partnered with Fortune 500 companies to drive AI-led transformation and business innovation. With a deep background in data and AI-driven strategy, Ashwin has worked across industries, including consumer goods, technology, financial services, and healthcare, helping organizations harness the power of advanced analytics, artificial intelligence, and machine learning.

Ashwin is renowned for his expertise in bridging the gap between cutting-edge technology and business impact. He has led the design and implementation of AI initiatives that have transformed decision-making, unlocked operational efficiencies, and delivered measurable financial results. Over his career, he has guided organizations through complex challenges such as scaling AI and analytics pilots to enterprise-wide deployments, realization of business value from AI initiatives, and driving cultural change to support AI adoption.

Ashwin's practical, hands-on experience informs the development of the AI Impact Model (AIIM). As a practitioner who has seen both the triumphs and challenges of AI adoption first-hand, Ashwin brings a unique perspective to the book, ensuring that its insights are grounded in real-world execution and tailored to the needs of business leaders.

Prof. Mohanbir Sawhney

Prof. Mohanbir Sawhney is the McCormick Foundation Professor of Technology and the Director of the Center for Research in Technology & Innovation at the Kellogg School of Management, Northwestern University. Over the past thirty-five years, Mohan has established himself as one of the foremost authorities on technology, innovation, and AI strategy. His research and consulting work focus on helping organizations harness technological advancements to drive growth, innovation, and transformation.

Mohan's work spans a diverse range of industries, including technology, consumer goods, manufacturing, and financial services. He has advised leading companies like Microsoft, Reliance Jio, Adobe, and Salesforce, as well as startups and emerging businesses, on navigating the complexities of AI and digital transformation. His contributions to the AI space include research on AI-driven decision-making, customer insights, and digital ecosystems, as well as his involvement in shaping AI strategies for major corporations.

A prolific writer, Mohan has authored seven management books and dozens of articles, including those featured in *Harvard Business Review* and *MIT Sloan Management Review*. He is a sought-after speaker and thought leader, recognized for his ability to bridge the worlds of academia and business with clarity and insight.

References

1. Challapally, A., Pease, C., Raskar, R., & Chari, P. (2025, July). *State of AI in Business 2025: The GenAI Divide*. MIT Nanda. https://mlq.ai/media/quarterly_decks/v0.1_State_of_AI_in_Business_2025_Report.pdf

2. KPMG LLP. (2024, October 21). *KPMG LLP AI & Digital Innovation Quarterly Pulse Survey*. kpmg.com. https://kpmg.com/us/en/media/news/digital-innovation-quarterly-pulse-survey.html

3. Perfony. (n.d.). *Successes and Failures of AI Projects*. https://www.perfony.com/en/the-successes-and-failures-of-ai-projects/

4. Uren, V. (n.d.). *Critical Success Factors for Artificial Intelligence Projects*. https://publications.aston.ac.uk/id/eprint/41673/1/euroma2020_full_paper_CSF_AI_1.5.pdf

5. Miller, G. J. (2022, March 22). *Artificial Intelligence Project Success Factors–Beyond the Ethical Principles*. https://link.springer.com/chapter/10.1007/978-3-030-98997-2_4

6. Gartner. (2021, July 14). *How to Improve Your Data Quality*. https:// www.gartner.com/smarterwithgartner/how-to-improve-your-data- quality

7. Redman, T. C. (2017, November 27). Seizing Opportunity in Data Quality. *MIT Sloan Management Review*. https://sloanreview.mit.edu/article/seizing-opportunity-in-data-quality/

8. KLEIN, J. Z. (2014, July 7). World Cup 2014: Germany Defeats Brazil, 7-1. *The New York Times*. https://www.nytimes.com/interactive/2014/07/08/sports/worldcup/world-cup-brazil-vs-germany.html

9. Ratzesberger, O., & Sawhney, M. (2017). *The Sentient Enterprise: The Evolution of Business Decision Making*. Wiley. https://www.amazon.com/Sentient-Enterprise-Evolution-Business-Decision/dp/1119438861

www.ingramcontent.com/pod-product-compliance
Ingram Content Group UK Ltd.
Pitfield, Milton Keynes, MK11 3LW, UK
UKHW021432280726
14060UKWH00001BA/46